The Student's Guide to VHDL

The Student's Guide to VHDL

Peter J. Ashenden
THE UNIVERSITY OF ADELAIDE

Morgan Kaufmann Publishers, Inc.
San Francisco, California

Senior Editor	Denise Penrose
Production Manager	Yonie Overton
Production Editor	Edward Wade
Production Assistant	Pamela Sullivan
Editorial Assistant	Meghan Keeffe
Cover Design	Ross Carron Design
Text Design	Rebecca Evans & Associates
Copyeditor	Ken DellaPenta
Printer	Courier Corporation

This book was author-typeset using Interleaf.

Appendices A and B reprinted from IEEE Std 1076-1993 "IEEE Standard VHDL Language Reference Manual," Copyright © 1994 and IEEE Std 1164-1993 "IEEE Standard Multivalue Logic System for VHDL Model Interoperability (Std_logic_1164)," Copyright © 1993 by the Institute of Electrical and Electronic Engineers, Inc. The IEEE disclaims any responsibility or liability resulting from the placement and use in the described manner. Information is reprinted with the permission of the IEEE.

Morgan Kaufmann Publishers, Inc.
Editorial and Sales Office
340 Pine Street, Sixth Floor
San Francisco, CA 94104-3205
USA

Telephone	415-392-2665
Facsimile	415-982-2665
Email	mkp@mkp.com
Website	*www.mkp.com*

Order toll free   800-745-7323

Library of Congress Cataloging-in-Publication Data

Ashenden, Peter J.
      The student's guide to VHDL / Peter J. Ashenden.
            p.   cm.
      Includes bibliographical references and index.
            ISBN 1–55860–520–7
      1. VHDL (Computer hardware description language) 2. Electronic
digital computers—Design and construction—Data processing.
      I. Title.
      TK7885.7.A84 1998
      621.39'2--dc21
                                                                    97-43565
                                                                      CIP

To my son Alexander

Contents

Preface

VHDL is a language for describing digital electronic systems. VHDL is designed to fill a number of needs in the design process. First, it allows description of the structure of a system, that is, how it is decomposed into subsystems and how those subsystems are interconnected. Second, it allows the specification of the function of a system using familiar programming language forms. Third, as a result, it allows the design of a system to be simulated before being manufactured, so that designers can quickly compare alternatives and test for correctness without the delay and expense of hardware prototyping. Fourth, it allows the detailed structure of a design to be synthesized from a more abstract specification, allowing designers to concentrate on more strategic design decisions and reducing time to market.

The Student's Guide to VHDL teaches the fundamental modeling features of VHDL, showing how the features are used for the design of digital systems. Features covered include the use of signals, entities and architectures for structural modeling; processes and programming language-like data types and statements for behavioral modeling; and concurrent statements for functional modeling. The book also covers some advanced features, including resolved signal types, generics and configuration, allowing students to tackle more sophisticated design projects.

The book is organized as a structured guide to VHDL, introducing students to the language through self-learning. As each language feature is introduced, examples show how to use the feature in the context of real VHDL models. Exercises at the end of each chapter test understanding and can be used for homework or lab exercises. Answers to the quiz-style exercises are included in Appendix E. Students will find some basic experience in a programming language, such as may be obtained from a freshman introduction to programming, helpful as preparation for use of this book.

The Student's Guide to VHDL is suitable for use as a textbook for introductory VHDL classes. The structured self-teaching approach also makes it suitable as a supplemental guide for courses that incorporate VHDL-based project work, for example, courses in digital systems, computer organization and computer architecture. The book will also serve practicing engineers who need to acquire basic VHDL fluency quickly, without going into the full details of the language.

The book does not purport to teach digital design, since that topic is large enough by itself to warrant several textbooks covering its various aspects. While *The Student's Guide to VHDL* does cover a large part of the language, there are some advanced topics that are not included. Those who need a comprehensive guide and reference for the whole language should read my book *The Designer's Guide to VHDL*. It covers all aspects of the language, illustrating them with examples and four fully worked case studies.

Structure of the Book

The Student's Guide to VHDL is organized so that it can be read linearly from front to back. This path offers a graduated development, with each chapter building on ideas introduced in the preceding chapters. Each chapter introduces a number of related concepts or language facilities and illustrates each one with examples.

Chapter 1 introduces the idea of a hardware description language and outlines the reasons for its use and the benefits that ensue. It then proceeds to introduce the basic concepts underlying VHDL, so that they can serve as a basis for examples in subsequent chapters. Readers with some previous introductory-level background in VHDL should treat Chapter 1 as review. The next three chapters cover the aspects of VHDL that are most like conventional programming languages. These may be used to describe the behavior of a system in algorithmic terms. Chapter 2 explains the basic type system of the language and introduces the scalar data types. Chapter 3 describes the sequential control structures, and Chapter 4 covers composite data structures used to represent collections of data elements. Readers who are proficient in using conventional programming languages should treat Chapters 2 through 4 as review.

In Chapter 5, the main facilities of VHDL used for modeling hardware are covered in detail. These include facilities for modeling the basic behavioral elements in a design, the signals that interconnect them and the hierarchical structure of the design. The combination of facilities described in these early chapters is sufficient for many modeling tasks.

The next two chapters extend this basic set of facilities with language features that make modeling of large systems more tractable. Chapter 6 introduces procedures and functions, which can be used to encapsulate behavioral aspects of a design. Chapter 7 introduces the package as a means of collecting together related parts of a design or of creating modules that can be reused in a number of designs.

The remaining chapters cover some of the advanced modeling features in VHDL. Chapter 8 deals with the important topic of resolved signals, and Chapter 9 describes generic constants as a means of parameterizing the behavior and structure of a design. While these language facilities form the basis of many real-world models, their treatment in this book is left to this late chapter. Experience has shown that the ideas can be difficult to understand without a solid foundation in the more basic language aspects. Chapter 10 deals with the topics of component instantiation and configuration. These features are also important in large real-world models, but they can be difficult to understand. Hence this book introduces structural modeling through the mechanism of direct instantiation in earlier chapters and leaves the more general case of component instantiation and configuration until this later chapter.

This book describes the 1993 version of VHDL, often referred to as VHDL-93. In those cases where VHDL-93 differs from the previous version of the language, VHDL-87, the book describes the differences and shows how to work around them. As an aid to users of software tools that only support VHDL-87, Appendix D summarizes the differences in one place.

Each chapter in the book is followed by a set of exercises designed to help the reader develop understanding of the material. Where an exercise relates to a particular topic described in the chapter, the section number is included in square brackets. An approximate "difficulty" rating is also provided, expressed using the following symbols:

❶   quiz-style exercise, testing basic understanding

❷   basic modeling exercise—10 minutes to half an hour effort

❸   advanced modeling exercise—half to two hours effort

Answers for the first category of exercises are provided in Appendix E. The remaining categories involve developing VHDL models. Readers are encouraged to test correctness of their models by running them on a VHDL simulator. This is a much more effective learning exercise than comparing paper models with paper solutions.

One pervasive theme running through the presentation in this book is that modeling a system using a hardware description language is essentially a software design exercise. This implies that good software engineering practice should be applied. Hence the treatment in this book draws directly from experience in software engineering. There are numerous hints and techniques from small-scale and large-scale software engineering presented throughout the book, with the sincere intention that they might be of use to readers.

The VHDL Standard

VHDL arose out of the United States government's Very High Speed Integrated Circuits (VHSIC) program. In the course of this program, it became clear that there was a need for a standard language for describing the structure and function of integrated circuits (ICs). Hence the VHSIC Hardware Description Language (VHDL) was developed. It was subsequently developed further under the auspices of the Institute of Electrical and Electronic Engineers (IEEE) and adopted in the form of the IEEE Standard 1076 *Standard VHDL Language Reference Manual* in 1987.

Like all IEEE standards, the VHDL standard is subject to review every five years. Comments and suggestions from users of the 1987 standard were analyzed by the IEEE working group responsible for VHDL, and in 1992 a revised version of the standard was proposed. This was eventually adopted in 1993. Work is currently in progress to update the standard for 1998.

The IEEE Standard 1076-1993 *Standard VHDL Language Reference Manual* is sometimes referred to as the "VHDL Bible." It is the authoritative source of information about VHDL. However, since it is a definitional document, not a tutorial, it is written in a complex legalistic style. This makes it very difficult to use to answer the usual questions that arise when writing VHDL models. It should only be used once you are somewhat familiar with VHDL. The *Language Reference Manual* is published by the IEEE. It can be ordered from the IEEE at the following address, quoting the publication number SH16840:

Ask*IEEE
Attn: Manager, Document Delivery
75 Varick Street, 9th Floor
New York, NY 10013

Phone: 800-949-IEEE (within US and Canada)
         212-301-4100 (outside US and Canada)

Fax: 212-301-4090 or 212-301-4091

E-mail: askieee@ieee.org

Resources for Help and Information

While this book covers many of the features of VHDL, there will no doubt be questions that it does not answer. For these, the reader will need to seek other resources.

Readers who have access to the Usenet electronic news network will find the news group comp.lang.vhdl a valuable resource. This discussion group is a source of announcements, sample models, questions and answers and useful software. Participants include VHDL users and people actively involved in the language standard working group and in VHDL tool development. The "frequently asked questions" (FAQ) file for this group is a mine of useful pointers to books, products and other information. It is archived at

http://vhdl.org/comp.lang.vhdl/

This book contains numerous examples of VHDL models that may also serve as a resource for resolving questions. The VHDL source code for these examples, as well as other resources and related information, is available for on-line access on the World Wide Web, using the URL

http://www.mkp.com/books_catalog/1-55860-520-7.asp

Although I have been careful to avoid errors in the example code, there are no doubt some that I have missed. I would be pleased to hear about them, so that I can correct them in the on-line code and in future printings of this book. Errata and general comments can be e-mailed to me at

petera@cs.adelaide.edu.au

Acknowledgements

The seeds for *The Designer's Guide to VHDL* go back to 1990 when I developed a brief set of notes, *The VHDL Cookbook*, for my computer architecture class at the University of Adelaide. At the time, there were few books on VHDL available, so I made my booklet available for on-line access. News of its availability spread quickly around the world, and within days, my e-mail in-box was bursting. At the time of writing this, over seven years later, I still regularly receive messages about the *Cookbook*. Many of the respondents urged me to write a full textbook version. With that encouragement, I embarked upon the exercise that led to *The Designer's Guide to VHDL*. I am grateful to the many engineers, students and teachers around the world who gave me that impetus.

Now, two years after publication of *The Designer's Guide*, the need for a book specifically for students has become evident. While *The Designer's Guide* is appropriate for those needing to learn VHDL in all its intricacies, it is too complete for use as a supplemental reference for many courses. I hope this book will meet the needs of those who need to learn just enough VHDL for the project work related to their courses.

When I embarked upon writing *The Designer's Guide*, I was not aware of the amount of effort that would be required. Thanks are due to my colleagues at the University of Adelaide for advice and support. Thanks also to Philip Wilsey and his postgraduate students at the University of Cincinnati for providing a lively and stimulating working environment during my sabbatical visit there in 1993. The book came along in leaps and bounds during that time.

Given the number of example models in this book, it was important that they be thoroughly checked. Model Technology, Inc., kindly donated use of their V-System analyzer and simulator for this purpose. I gratefully acknowledge this contribution and thank them for their timely responses to my frequent questions.

In the preface to *The Designer's Guide to VHDL*, I extended thanks to numerous people, in particular, the staff at Morgan Kaufmann, and the reviewers of that book: Poras Balsara, Paul Menchini, David Pitts and Philip Wilsey. Naturally my gratitude continues. For *The Student's Guide to VHDL*, I again extend thanks to the Morgan Kaufmann staff, especially Denise Penrose for her continued enthusiasm and sound advice, and to the reviewers who commented on the selection of material: Venkatesh Akella, Manoj Franklin, Kenneth J. Hintz, David Kaeli, Chris Myers, Mike Smith, Jerry Sobelman and Edward Stabler.

I dedicated *The Designer's Guide* to my wife, Katrina, in appreciation of her understanding, encouragement and support during the writing of that book. We now have a little boy, Alexander, whose boundless energy and constant curiosity are a real joy. I dedicate this book to him, in the hope that his spirit of wonder will stay with him as he grows.

Fundamental Concepts 1

In this introductory chapter, we describe what we mean by digital system modeling and see why modeling and simulation are an important part of the design process. We see how the hardware description language VHDL can be used to model digital systems, and introduce some of the basic concepts underlying the language. We complete this chapter with a description of the basic lexical and syntactic elements of the language, to form a basis for the detailed descriptions of language features that follow in later chapters.

1.1 Modeling Digital Systems

If we are to discuss the topic of modeling digital systems, we first need to agree on what a digital system is. Different engineers would come up with different definitions, depending on their background and the field in which they were working. Some may consider a single VLSI circuit to be a self-contained digital system. Others might take a larger view and think of a complete computer, packaged in a cabinet with peripheral controllers and other interfaces.

For the purposes of this book, we include any digital circuit that processes or stores information as a digital system. We thus consider both the system as a whole and the various parts from which it is constructed. Thus our discussions cover a range of systems from the low-level gates that make up the components to the top-level functional units.

If we are to encompass this range of views of digital systems, we must recognize the complexity with which we are dealing. It is not humanly possible to comprehend such complex systems in their entirety. We need to find methods of dealing with the complexity, so that we can, with some degree of confidence, design components and systems that meet their requirements.

The most important way of meeting this challenge is to adopt a systematic methodology of design. If we start with a requirements document for the system, we can design an abstract structure that meets the requirements. We can then decompose this structure into a collection of components that interact to perform the same function. Each of these components can in turn be decomposed until we get to a level where we have some ready-made, primitive components that perform a required function. The result of this process is a hierarchically composed system, built from the primitive elements.

The advantage of this methodology is that each subsystem can be designed independently of others. When we use a subsystem, we can think of it as an abstraction rather than having to consider its detailed composition. So at any particular stage in the design process, we only need to pay attention to the small amount of information relevant to the current focus of design. We are saved from being overwhelmed by masses of detail.

We use the term *model* to mean our understanding of a system. The model represents that information which is relevant and abstracts away from irrelevant detail. The implication of this is that there may be several models of the same system, since different information is relevant in different contexts. One kind of model might concentrate on representing the function of the system, whereas another kind might represent the way in which the system is composed of subsystems. We will come back to this idea in more detail in the next section.

There are a number of important motivations for formalizing this idea of a model. First, when a digital system is needed, the requirements of the system must be specified. The job of the engineers is to design a system that meets these requirements. To do that, they must be given an understanding of the requirements, hopefully in a way that leaves them free to explore alternative implementations and to choose the best according to some criteria. One of the problems that often arises is that requirements are incompletely and ambiguously spelled out, and the customer and the design engineers disagree on what is meant by the requirements document. This problem can be avoided by using a formal model to communicate requirements.

A second reason for using formal models is to communicate understanding of the function of a system to a user. The designer cannot always predict every possible way in which a system may be used, and so is not able to enumerate all possible behaviors. If the designer provides a model, the user can check it against any given set of inputs and determine how the system behaves in that context. Thus a formal model is an invaluable tool for documenting a system.

A third motivation for modeling is to allow testing and verification of a design using simulation. If we start with a requirements model that defines the behavior of a system, we can simulate the behavior using test inputs and note the resultant outputs of the system. According to our design methodology, we can then design a circuit from subsystems, each with its own model of behavior. We can simulate this composite system with the same test inputs and compare the outputs with those of the previous simulation. If they are the same, we know that the composite system meets the requirements for the cases tested. Otherwise we know that some revision of the design is needed. We can continue this process until we reach the bottom level in our design hierarchy, where the components are real devices whose behavior we know. Subsequently, when the design is manufactured, the test inputs and outputs from simulation can be used to verify that the physical circuit functions correctly. This approach to testing and verification of course assumes that the test inputs cover all of the circumstances in which the final circuit will be used. The issue of test coverage is a complex problem in itself and is an active area of research.

A fourth motivation for modeling is to allow formal verification of the correctness of a design. Formal verification requires a mathematical statement of the required function of a system. This statement may be expressed in the notation of a formal logic system, such as temporal logic. Formal verification also requires a mathematical definition of the meaning of the modeling language or notation used to describe a design. The process of verification involves application of the rules of inference of the logic system to prove that the design implies the required function. While formal verification is not yet in everyday use, it is an active area of research. There have already been significant demonstrations of formal verification techniques in real design projects, and the promise for the future is bright.

One final, but equally important, motivation for modeling is to allow automatic synthesis of circuits. If we can formally specify the function required of a system, it is in theory possible to translate that specification into a circuit that performs the function. The advantage of this approach is that the human cost of design is reduced, and engineers are free to explore alternatives rather than being bogged down in design detail. Also, there is less scope for errors being introduced into a design and not being detected. If we automate the translation from specification to implementation, we can be more confident that the resulting circuit is correct.

The unifying factor behind all of these arguments is that we want to achieve maximum reliability in the design process for minimum cost and design time. We need to ensure that requirements are clearly specified and understood, that subsystems are used correctly and that designs meet the requirements. A major contributor to excessive cost is having to revise a design after manufacture to correct errors. By avoiding errors, and by providing better tools for the design process, costs and delays can be contained.

1.2   Domains and Levels of Modeling

In the previous section, we mentioned that there may be different models of a system, each focussing on different aspects. We can classify these models into three domains: *function, structure* and *geometry*. The functional domain is concerned with the operations performed by the system. In a sense, this is the most abstract domain of description, since it does not indicate how the function is implemented. The structural domain deals with how the system is composed of interconnected subsystems. The geometric domain deals with how the system is laid out in physical space.

Each of these domains can also be divided into levels of abstraction. At the top level, we consider an overview of function, structure or geometry, and at lower levels we introduce successively finer detail. Figure 1-1 (devised by Gajski and Kuhn, see reference [2]) represents the domains on three independent axes, and represents the levels of abstraction by the concentric circles crossing each of the axes.

FIGURE 1-1

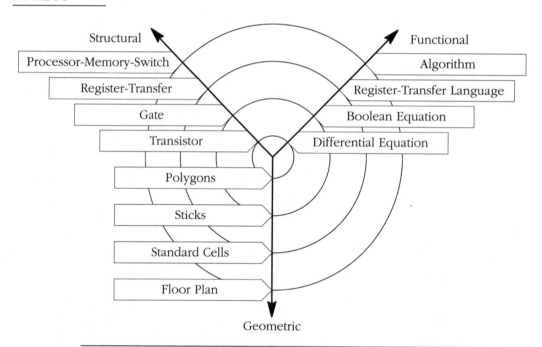

Domains and levels of abstraction. The radial axes show the three different domains of modeling. The concentric rings show the levels of abstraction, with the more abstract levels on the outside and more detailed levels towards the center.

Let us look at this classification in more detail, showing how at each level we can create models in each domain. As an example, we consider a single-chip microcontroller system used as the controller for some measurement instrument, with data input connections and some form of display outputs.

At the most abstract level, the function of the entire system may be described in terms of an algorithm, much like an algorithm for a computer program. This level of functional modeling is often called *behavioral modeling,* a term we shall adopt when presenting abstract descriptions of a system's function. A possible algorithm for our instrument controller is shown in Figure 1-2. This model describes how the controller repeatedly scans each data input and writes a scaled display of the input value.

FIGURE 1-2

```
loop
    for each data input loop
        read the value on this input;
        scale the value using the current scale factor for this input;
        convert the scaled value to a decimal string;
        write the string to the display output corresponding to this input;
    end loop;
    wait for 10 ms;
end loop;
```

An algorithm for a measurement instrument controller.

At this top level of abstraction, the structure of a system may be described as an interconnection of such components as processors, memories and input/output devices. This level is sometimes called the Processor Memory Switch (PMS) level, named after the notation used by Bell and Newell (see reference [1]). Figure 1-3 shows a structural model of the instrument controller drawn using this notation. It consists of a processor connected via a switch to a memory component and to controllers for the data inputs and display outputs.

FIGURE 1-3

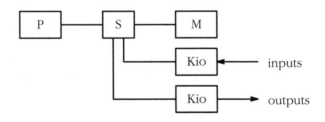

A PMS model of the controller structure. It is constructed from a processor (P), a memory (M), an interconnection switch (S) and two input/output controllers (Kio).

In the geometric domain at this top level of abstraction, a system to be implemented as a VLSI circuit may be modeled using a floor plan. This shows how the components described in the structural model are arranged on the silicon die. Figure 1-4 shows a possible floor plan for the instrument controller chip. There are analogous geometric descriptions for systems integrated in other media. For example, a personal computer system might be modeled at the top level in the geometric domain by an assembly diagram showing the positions of the motherboard and plug-in expansion boards in the desktop cabinet.

The next level of abstraction in modeling, depicted by the second ring in Figure 1-1, describes the system in terms of units of data storage and transformation. In the structur-

FIGURE 1-4

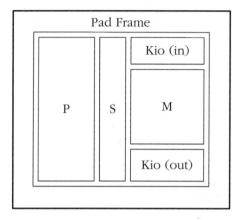

A floor plan model of the controller geometry.

FIGURE 1-5

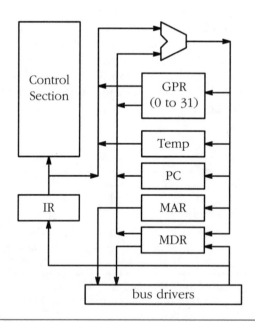

A register-transfer-level structural model of the controller processor. It consists of a general purpose register (GPR) file, registers for the program counter (PC), memory address (MAR), memory data (MDR), temporary values (Temp) and fetched instructions (IR), an arithmetic unit, bus drivers and the control section.

al domain, this is often called the *register-transfer* level, composed of a *data path* and a *control section*. The data path contains data storage registers, and data is transferred between them through transformation units. The control section sequences operation of the data path components. For example, a register-transfer-level structural model of the processor in our controller is shown in Figure 1-5.

In the functional domain, a *register-transfer language* (RTL) is often used to specify the operation of a system at this level. Storage of data is represented using register variables, and transformations are represented by arithmetic and logical operators. For example, an RTL model for the processor in our example controller might include the following description:

```
MAR ← PC, memory_read ← 1
PC ← PC + 1
wait until ready = 1
IR ← memory_data
memory_read ← 0
```

This section of the model describes the operations involved in fetching an instruction from memory. The contents of the PC register are transferred to the memory address register, and the **memory_read** signal is asserted. Then the value from the PC register is transformed (incremented in this case) and transferred back to the PC register. When the **ready** input from the memory is asserted, the value on the memory data input is transferred to the instruction register. Finally, the **memory_read** signal is negated.

In the geometric domain, the kind of model used depends on the physical medium. In our example, standard library cells might be used to implement the registers and data transformation units, and these must be placed in the areas allocated in the chip floor plan.

The third level of abstraction shown in Figure 1-1 is the conventional logic level. At this level, structure is modeled using interconnections of gates, and function is modeled by Boolean equations or truth tables. In the physical medium of a custom integrated circuit, geometry may be modeled using a virtual grid, or "sticks," notation.

At the most detailed level of abstraction, we can model structure using individual transistors, function using the differential equations that relate voltage and current in the circuit, and geometry using polygons for each mask layer of an integrated circuit. Most designers do not need to work at this detailed level, as design tools are available to automate translation from a higher level.

1.3   Modeling Languages

In the previous section, we saw that different kinds of models can be devised to represent the various levels of function, structure and physical arrangement of a system. There are also different ways of expressing these models, depending on the use made of the model.

As an example, consider the ways in which a structural model may be expressed. One common form is a circuit schematic. Graphical symbols are used to represent subsystems, and instances of these are connected using lines that represent wires. This graphical form is generally the one preferred by designers. However, the same structural information can be represented textually in the form of a net list.

When we move into the functional domain, we usually see textual notations used for modeling. Some of these are intended for use as specification languages, to meet the need for describing the operation of a system without indicating how it might be implemented. These notations are usually based on formal mathematical methods, such as temporal logic or abstract state machines. Other notations are intended for simulating the system for test and verification purposes and are typically based on conventional programming languages. Yet other notations are oriented towards hardware synthesis and usually have a more restricted set of modeling facilities, since some programming language constructs are difficult to translate into hardware.

The purpose of this book is to describe the modeling language VHDL. VHDL includes facilities for describing structure and function at a number of levels, from the most abstract down to the gate level. It also provides an attribute mechanism that can be used to annotate a model with information in the geometric domain. VHDL is intended, among other things, as a modeling language for specification and simulation. We can also use it for hardware synthesis if we restrict ourselves to a subset that can be automatically translated into hardware.

1.4   VHDL Modeling Concepts

In the previous section, we looked at the three domains of modeling: function, structure and geometry. In this section, we look at the basic modeling concepts in each of these domains and introduce the corresponding VHDL elements for describing them. This will provide a feel for VHDL and a basis from which to work in later chapters. As an example, we look at ways of describing a four-bit register, shown in Figure 1-6.

FIGURE 1-6

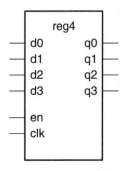

A four-bit register module. The register is named reg4 *and has six inputs,* d0, d1, d2, d3, en *and* clk, *and four outputs,* q0, q1, q2 *and* q3.

FIGURE 1-7

```
entity reg4 is
    port ( d0, d1, d2, d3, en, clk : in bit;
           q0, q1, q2, q3 : out  bit );
end entity reg4;
```

A VHDL entity description of a four-bit register.

Using VHDL terminology, we call the module **reg4** a design *entity*, and the inputs and outputs are *ports*. Figure 1-7 shows a VHDL description of the interface to this entity. This is an example of an *entity declaration*. It introduces a name for the entity and lists the input and output ports, specifying that they carry bit values ('0' or '1') into and out of the entity. From this we see that an entity declaration describes the external view of the entity.

Elements of Behavior

In VHDL, a description of the internal implementation of an entity is called an *architecture body* of the entity. There may be a number of different architecture bodies of the one interface to an entity, corresponding to alternative implementations that perform the same function. We can write a *behavioral* architecture body of an entity, which describes the function in an abstract way. Such an architecture body includes only *process statements*, which are collections of actions to be executed in sequence. These actions are called *sequential statements* and are much like the kinds of statements we see in a conventional programming language. The types of actions that can be performed include evaluating expressions, assigning values to variables, conditional execution, repeated execution and subprogram calls. In addition, there is a sequential statement that is unique to hardware modeling languages, the *signal assignment* statement. This is similar to variable assignment, except that it causes the value on a signal to be updated at some future time.

To illustrate these ideas, let us look at a behavioral architecture body for the reg4 entity, shown in Figure 1-8. In this architecture body, the part after the first **begin** keyword includes one process statement, which describes how the register behaves. It starts with the process name, storage, and finishes with the keywords **end process**.

The process statement defines a sequence of actions that are to take place when the system is simulated. These actions control how the values on the entity's ports change over time, that is, they control the behavior of the entity. This process can modify the values of the entity's ports using signal assignment statements.

FIGURE 1-8

```
architecture behav of reg4 is
begin
    storage : process is
        variable stored_d0, stored_d1, stored_d2, stored_d3 : bit;
    begin
        if en = '1' and clk = '1' then
            stored_d0 := d0;
            stored_d1 := d1;
            stored_d2 := d2;
            stored_d3 := d3;
        end if;
        q0 <= stored_d0 after 5 ns;
        q1 <= stored_d1 after 5 ns;
        q2 <= stored_d2 after 5 ns;
        q3 <= stored_d3 after 5 ns;
        wait on d0, d1, d2, d3, en, clk;
    end process storage;
end architecture behav;
```

A behavioral architecture body of the reg4 *entity.*

The way this process works is as follows. When the simulation is started, the signal values are set to '0', and the process is activated. The process's variables (listed after the keyword **variable**) are initialized to '0', then the statements are executed in order. The first statement is a condition that tests whether the values of the en and clk signals are both '1'. If they are, the statements between the keywords **then** and **end if** are executed, updating the process's variables using the values on the input signals. After the conditional if statement, there are four signal assignment statements that cause the output signals to be updated 5 ns later.

When all of these statements in the process have been executed, the process reaches the *wait statement* and *suspends*, that is, it becomes inactive. It stays suspended until one of the signals to which it is *sensitive* changes value. In this case, the process is sensitive to the signals d0, d1, d2, d3, en and clk, since they are listed in the wait statement. When one of these changes value, the process is resumed. The statements are executed again, starting from the keyword **begin**, and the cycle repeats. Notice that while the process is suspended, the values in the process's variables are not lost. This is how the process can represent the state of a system.

Elements of Structure

An alternative way of describing the implementation of an entity is to specify how it is composed of subsystems. We can give a structural description of the entity's implementation. An architecture body that is composed only of interconnected subsystems is called a *structural* architecture body. Figure 1-9 shows how the reg4 entity might be composed of latches and gates. If we are to describe this in VHDL, we will need entity declarations and architecture bodies for the subsystems, shown in Figure 1-10.

Figure 1-11 is a VHDL architecture body declaration that describes the structure shown in Figure 1-9. The *signal declaration*, before the keyword **begin**, defines the internal signals of the architecture. In this example, the signal int_clk is declared to carry a bit value ('0' or '1'). In general, VHDL signals can be declared to carry arbitrarily complex values. Within the architecture body the ports of the entity are also treated as signals.

In the second part of the architecture body, a number of *component instances* are created, representing the subsystems from which the reg4 entity is composed. Each component instance is a copy of the entity representing the subsystem, using the corresponding basic architecture body. (The name work refers to the current working library, in which all of the entity and architecture body descriptions are assumed to be held.)

The *port map* specifies the connection of the ports of each component instance to signals within the enclosing architecture body. For example, bit0, an instance of the d_latch entity, has its port d connected to the signal d0, its port clk connected to the signal int_clk and its port q connected to the signal q0.

Mixed Structural and Behavioral Models

Models need not be purely structural or purely behavioral. Often it is useful to specify a model with some parts composed of interconnected component instances, and other parts described using processes. We use signals as the means of joining component instances and processes. A signal can be associated with a port of a component instance and can also be assigned to or read in a process.

FIGURE 1-9

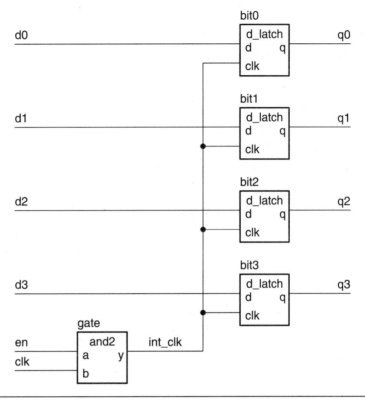

A structural composition of the reg4 *entity.*

FIGURE 1-10

```
entity d_latch is
    port ( d, clk : in bit;  q : out bit );
end d_latch;

architecture basic of d_latch is
begin

    latch_behavior : process is
    begin
        if clk = '1' then
            q <= d after 2 ns;
        end if;
        wait on clk, d;
    end process latch_behavior;

end architecture basic;
```

(continued on page 12)

(continued from page 11)

```
entity and2 is
    port ( a, b : in bit;  y : out bit );
end and2;

architecture basic of and2 is
begin

    and2_behavior : process is
    begin
        y <= a and b after 2 ns;
        wait on a, b;
    end process and2_behavior;

end architecture basic;
```

Entity declarations and architecture bodies for D-flipflop and two-input and gate.

FIGURE 1-11

```
architecture struct of reg4 is

    signal int_clk : bit;

begin

    bit0 : entity work.d_latch(basic)
        port map (d0, int_clk, q0);
    bit1 : entity work.d_latch(basic)
        port map (d1, int_clk, q1);
    bit2 : entity work.d_latch(basic)
        port map (d2, int_clk, q2);
    bit3 : entity work.d_latch(basic)
        port map (d3, int_clk, q3);

    gate : entity work.and2(basic)
        port map (en, clk, int_clk);

end architecture struct;
```

A VHDL structural architecture body of the **reg4** *entity.*

We can write such a hybrid model by including both component instance and process statements in the body of an architecture. These statements are collectively called *concurrent statements*, since the corresponding processes all execute concurrently when the model is simulated. An outline of such a model is shown in Figure 1-12. This model describes a multiplier consisting of a data path and a control section. The data path is described structurally, using a number of component instances. The control section is described behaviorally, using a process that assigns to the control signals for the data path.

Test Benches

In our introductory discussion, we mentioned testing through simulation as an important motivation for modeling. We often test a VHDL model using an enclosing model

FIGURE 1-12

```
entity multiplier is
    port ( clk, reset : in bit;
            multiplicand, multiplier : in integer;
            product : out integer );
end entity multiplier;
```

- -

```
architecture mixed of multiplier is
    signal partial_product, full_product : integer;
    signal arith_control, result_en, mult_bit, mult_load : bit;
begin -- mixed
    arith_unit : entity work.shift_adder(behavior)
        port map ( addend => multiplicand, augend => full_product,
                    sum => partial_product,
                    add_control => arith_control);
    result : entity work.reg(behavior)
        port map ( d => partial_product, q => full_product,
                    en => result_en, reset => reset);
    multiplier_sr : entity work.shift_reg(behavior)
        port map ( d => multiplier, q => mult_bit,
                    load => mult_load, clk => clk);
    product <= full_product;
    control_section : process is
        -- variable declarations for control_section
        -- . . .
    begin -- control section
        -- sequential statements to assign values to control signals
        -- . . .
        wait on clk, reset;
    end process control_section;
end architecture mixed;
```

An outline of a mixed structural and behavioral model of a multiplier.

called a *test bench*. The name comes from the analogy with a real hardware test bench, on which a device under test is stimulated with signal generators and observed with signal probes. A VHDL test bench consists of an architecture body containing an instance of the component to be tested and processes that generate sequences of values on signals connected to the component instance. The architecture body may also contain processes that test that the component instance produces the expected values on its output signals. Alternatively, we may use the monitoring facilities of a simulator to observe the outputs.

A test bench model for the behavioral implementation of the **reg4** register is shown in Figure 1-13. The entity declaration has no port list, since the test bench is entirely self-contained. The architecture body contains signals that are connected to the input and output ports of the component instance **dut**, the device under test. The process labeled **stimulus** provides a sequence of test values on the input signals by performing signal assignment statements, interspersed with wait statements. Each wait statement specifies a 20 ns pause during which the register device determines its output values. We can use a simulator to observe the values on the signals **q0** to **q3** to verify that the register operates correctly. When all of the stimulus values have been applied, the stimulus process waits indefinitely, thus completing the simulation.

FIGURE 1-13

```
entity test_bench is
end entity test_bench;

- - - - - - - - - - - - - - - - - - - - - - - - - - - - - - - - - - - - - - -

architecture test_reg4 of test_bench is
    signal d0, d1, d2, d3, en, clk, q0, q1, q2, q3 : bit;
begin
    dut : entity work.reg4(behav)
        port map ( d0, d1, d2, d3, en, clk, q0, q1, q2, q3 );
    stimulus : process is
    begin
        d0 <= '1';  d1 <= '1';  d2 <= '1';  d3 <= '1';
        en <= '0';  clk <= '0';
        wait for 20 ns;
        en <= '1';  wait for 20 ns;
        clk <= '1';  wait for 20 ns;
        d0 <= '0';  d1 <= '0';  d2 <= '0';  d3 <= '0';  wait for 20 ns;
        en <= '0';  wait for 20 ns;
        . . .
        wait;
    end process stimulus;
end architecture test_reg4;
```

A VHDL test bench for the **reg4** *register model.*

Analysis, Elaboration and Execution

One of the main reasons for writing a model of a system is to enable us to simulate it. This involves three stages: *analysis, elaboration* and *execution*. Analysis and elaboration are also required in preparation for other uses of the model, such as logic synthesis.

In the first stage, analysis, the VHDL description of a system is checked for various kinds of errors. Like most programming languages, VHDL has rigidly defined *syntax* and *semantics*. The syntax is the set of grammatical rules that govern how a model is written. The rules of semantics govern the meaning of a program. For example, it makes sense to perform an addition operation on two numbers but not on two processes.

During the analysis phase, the VHDL description is examined, and syntactic and static semantic errors are located. The whole model of a system need not be analyzed at once. Instead, it is possible to analyze *design units*, such as entity and architecture body declarations, separately. If the analyzer finds no errors in a design unit, it creates an intermediate representation of the unit and stores it in a library. The exact mechanism varies between VHDL tools.

The second stage in simulating a model, elaboration, is the act of working through the design hierarchy and creating all of the objects defined in declarations. The ultimate product of design elaboration is a collection of signals and processes, with each process possibly containing variables. A model must be reducible to a collection of signals and processes in order to simulate it.

We can see how elaboration achieves this reduction by starting at the top level of a model, namely, an entity, and choosing an architecture of the entity to simulate. The architecture comprises signals, processes and component instances. Each component instance is a copy of an entity and an architecture that also comprises signals, processes and component instances. Instances of those signals and processes are created, corresponding to the component instance, and then the elaboration operation is repeated for the sub-component instances. Ultimately, a component instance is reached that is a copy of an entity with a purely behavioral architecture, containing only processes. This corresponds to a primitive component for the level of design being simulated. Figure 1-14 shows how elaboration proceeds for the structural architecture body of the **reg4** entity. As each instance of a process is created, its variables are created and given initial values. We can think of each process instance as corresponding to one instance of a component.

The third stage of simulation is the execution of the model. The passage of time is simulated in discrete steps, depending on when events occur. Hence the term *discrete event simulation* is used. At some simulation time, a process may be stimulated by changing the value on a signal to which it is sensitive. The process is resumed and may schedule new values to be given to signals at some later simulated time. This is called *scheduling a transaction* on that signal. If the new value is different from the previous value on the signal, an *event* occurs, and other processes sensitive to the signal may be resumed.

The simulation starts with an *initialization phase*, followed by repetitive execution of a *simulation cycle*. During the initialization phase, each signal is given an initial value, depending on its type. The simulation time is set to zero, then each process instance is activated and its sequential statements executed. Usually, a process will include a signal assignment statement to schedule a transaction on a signal at some later simulation time. Execution of a process continues until it reaches a wait statement, which causes the process to be suspended.

During the simulation cycle, the simulation time is first advanced to the next time at which a transaction on a signal has been scheduled. Second, all the transactions scheduled for that time are performed. This may cause some events to occur on some signals. Third, all processes that are sensitive to those events are resumed and are allowed to continue until they reach a wait statement and suspend. Again, the processes usually execute signal assignments to schedule further transactions on signals. When all the processes have suspended again, the simulation cycle is repeated. When the simulation gets to the stage where there are no further transactions scheduled, it stops, since the simulation is then complete.

FIGURE 1-14

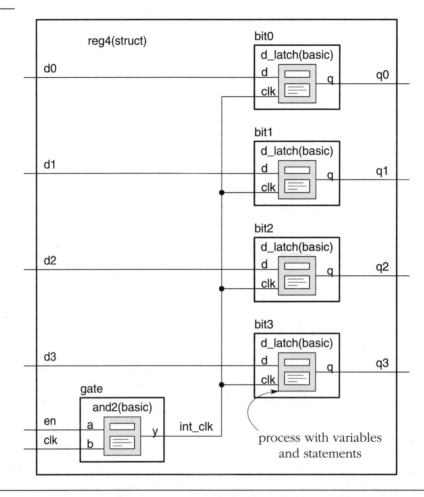

The elaboration of the **reg4** *entity using the structural architecture body. Each instance of the* **d_latch** *and* **and2** *entities is replaced with the contents of the corresponding* **basic** *architecture. These each consist of a process with its variables and statements.*

1.5   Learning a New Language: Lexical Elements and Syntax

When we learn a new natural language, such as Greek, Chinese or English, we start by learning the alphabet of symbols used in the language, then form these symbols into words. Next, we learn the way to put the words together to form sentences, and learn the meaning of these combinations of words. We reach fluency in a language when we can easily express what we need to say using correctly formed sentences.

    The same ideas apply when we need to learn a new special-purpose language, such as VHDL for describing digital systems. We can borrow a few terms from language theory to describe what we need to learn. First, we need to learn the alphabet with which the language is written. The VHDL alphabet consists of all of the characters in the ISO eight-bit character set. This includes uppercase and lowercase letters (including letters with diacritical marks, such as à, ä, and so forth), digits 0 to 9, punctuation and other special characters. Second, we need to learn the *lexical elements* of the language. In VHDL, these are the identifiers, reserved words, special symbols and literals. Third, we need to learn the *syntax* of the language. This is the grammar that determines what combinations of lexical elements make up legal VHDL descriptions. Fourth, we need to learn the *semantics*, or meaning, of VHDL descriptions. It is the semantics that allow a collection of symbols to describe a digital design. Fifth, we need to learn how to develop our own VHDL descriptions to describe a design we are working with. This is the creative part of modeling, and fluency in this part will greatly enhance our design skills.

In the remainder of this chapter, we describe the lexical elements used in VHDL and introduce the notation we use to describe the syntax rules. Then in subsequent chapters, we introduce the different facilities available in the language. For each of these, we show the syntax rules, describe the corresponding semantics and give examples of how they are used to model particular parts of a digital system. We also include some exercises at the end of each chapter to provide practice in the fifth stage of learning described above.

VHDL-87

VHDL-87 uses the ASCII character set, rather than the full ISO character set. ASCII is a subset of the ISO character set, consisting of just the first 128 characters. This includes all of the unaccented letters, but excludes letters with diacritical marks.

Lexical Elements

In the following section, we discuss the lexical elements of VHDL: *comments, identifiers, reserved words, special symbols, numbers, characters, strings* and *bit strings*.

Comments

When we are writing a hardware model in VHDL, it is important to annotate the code with comments. The reason for doing this is to help readers understand the structure and logic behind the model. It is important to realize that although we only write a model once, it may subsequently be read and modified many times, both by its author and by other engineers. Any assistance we can give to understanding the model is worth the effort.

A VHDL model consists of a number of lines of text. A comment can be added to a line by writing two dashes together, followed by the comment text. For example:

> ... *a line of VHDL description* ...   – – *a descriptive comment*

The comment extends from the two dashes to the end of the line and may include any text we wish, since it is not formally part of the VHDL model. The code of a model can include blank lines and lines that only contain comments, starting with two dashes. We can write long comments on successive lines, each starting with two dashes, for example:

```
-- The following code models
-- the control section of the system
... some VHDL code ...
```

Identifiers

Identifiers are used to name items in a VHDL model. It is good practice to use names that indicate the purpose of the item, so VHDL allows names to be arbitrarily long. However, there are some rules about how identifiers may be formed. A basic identifier

- may only contain alphabetic letters ('A' to 'Z' and 'a' to 'z'), decimal digits ('0' to '9') and the underline character ('_');

- must start with an alphabetic letter;

- may not end with an underline character; and

- may not include two successive underline characters.

Some examples of valid basic identifiers are

A   X0   counter   Next_Value   generate_read_cycle

Some examples of invalid basic identifiers are

```
last@value        -- contains an illegal character for an identifier
5bit_counter      -- starts with a non-alphabetic character
_A0               -- starts with an underline
A0_               -- ends with an underline
clock__pulse      -- two successive underlines
```

Note that the case of letters is not considered significant, so the identifiers **cat** and **Cat** are the same. Underline characters in identifiers are significant, so **This_Name** and **ThisName** are different identifiers.

In addition to the basic identifiers, VHDL allows *extended identifiers*, which can contain any sequence of characters. Extended identifiers are included to allow communication between computer-aided engineering tools for processing VHDL descriptions and other tools that use different rules for identifiers. An extended identifier is written by enclosing the characters of the identifier between '\' characters. For example:

\data bus\   \global.clock\   \923\   \d#1\   \start_ _\

If we need to include a '\' character in an extended identifier, we do so by doubling the character, for example:

\A:\\name\   -- contains a '\' between the ':' and the 'n'

Note that the case of letters is significant in extended identifiers and that all extended identifiers are distinct from all basic identifiers. So the following are all distinct identifiers:

name   \name\   \Name\   \NAME\

VHDL-87

VHDL-87 only allows basic identifiers, not extended identifiers. The rules for forming basic identifiers are the same as those for VHDL-93.

Reserved Words

Some identifiers, called reserved words or keywords, are reserved for special use in VHDL. They are used to denote specific constructs that form a model, so we cannot use them as identifiers for items we define. The full list of reserved words is shown in Figure 1-15. Often, when a VHDL program is typeset, reserved words are printed in boldface. This convention is followed in this book.

FIGURE 1-15

abs	**entity**	**next**	**select**
access	**exit**	**nor**	**severity**
after	**file**	**not**	**signal**
alias	**for**	**null**	**shared**
all	**function**	**of**	**sla**
and	**generate**	**on**	**sll**
architecture	**generic**	**open**	**sra**
array	**group**	**or**	**srl**
assert	**guarded**	**others**	**subtype**
attribute	**if**	**out**	**then**
begin	**impure**	**package**	**to**
block	**in**	**port**	**transport**
body	**inertial**	**postponed**	**type**
buffer	**inout**	**procedure**	**unaffected**
bus	**is**	**process**	**units**
case	**label**	**pure**	**until**
component	**library**	**range**	**use**
configuration	**linkage**	**record**	**variable**
constant	**literal**	**register**	**wait**
disconnect	**loop**	**reject**	**when**
downto	**map**	**rem**	**while**
else	**mod**	**report**	**with**
elsif	**nand**	**return**	**xnor**
end	**new**	**rol**	**xor**
		ror	

VHDL reserved words.

VHDL-87

The following identifiers are not used as reserved words in VHDL-87. They may be used as identifiers for other purposes, although it is not advisable to do so, as this may cause difficulties in porting the models to VHDL-93.

group	postponed	ror	sra
impure	pure	shared	srl
inertial	reject	sla	unaffected
literal	rol	sll	xnor

Special Symbols

VHDL uses a number of special symbols to denote operators, to delimit parts of language constructs and as punctuation. Some of these special symbols consist of just one character. They are

& ' ( ) * + , − . / : ; < = > |

Other special symbols consist of pairs of characters. The two characters must be typed next to each other, with no intervening space. These symbols are

=> ** := /= >= <= <>

Numbers

There are two forms of numbers that can be written in VHDL code: *integer literals* and *real literals*. An integer literal simply represents a whole number and consists of digits without a decimal point. Real literals, on the other hand, can represent fractional numbers. They always include a decimal point, which is preceded by at least one digit and followed by at least one digit. Real literals represent an approximation to real numbers.

Some examples of decimal integer literals are

23   0   146

Note that **−10**, for example, is not an integer literal. It is actually a combination of a negation operator and the integer literal **10**.

Some examples of real literals are

23.1   0.0   3.14159

Both integer and real literals can also use exponential notation, in which the number is followed by the letter 'E' or 'e', and an exponent value. This indicates a power of 10 by which the number is multiplied. For integer literals, the exponent must not be negative, whereas for real literals, it may be either positive or negative. Some examples of integer literals using exponential notation are

46E5   1E+12   19e00

Some examples of real literals using exponential notation are

1.234E09   98.6E+21   34.0e−08

Integer and real literals may also be expressed in a base other than base 10. In fact, the base can be any integer between 2 and 16. To do this, we write the number surrounded by sharp characters ('#'), preceded by the base. For bases greater than 10, the letters 'A' through 'F' (or 'a' through 'f') are used to represent the digits 10 through 15. For example, several ways of writing the value 253 are as follows:

2#11111101#   16#FD#   16#0fd#   8#0375#

Similarly, the value 0.5 can be represented as

    2#0.100#   8#0.4#   12#0.6#

Note that in all these cases, the base itself is expressed in *decimal*.

Based literals can also use exponential notation. In this case, the exponent, expressed in decimal, is appended to the based number after the closing sharp character. The exponent represents the power of the base by which the number is multiplied. For example, the number 1024 could be represented by the integer literals:

    2#1#E10   16#4#E2   10#1024#E+00

Finally, as an aid to readability of long numbers, we can include underline characters as separators between digits. The rules for including underline characters are similar to those for identifiers; that is, they may not appear at the beginning or end of a number, nor may two appear in succession. Some examples are

    123_456   3.141_592_6   2#1111_1100_0000_0000#

Characters

A character literal can be written in VHDL code by enclosing it in single quotation marks. Any of the printable characters in the standard character set (including a space character) can be written in this way. Some examples are

    'A'   – – *uppercase letter*
    'z'   – – *lowercase letter*
    ','   – – *the punctuation character comma*
    '"'   – – *the punctuation character single quote*
    ' '   – – *the separator character space*

Strings

A string literal represents a sequence of characters and is written by enclosing the characters in double quotation marks. The string may include any number of characters (including zero), but it must fit entirely on one line. Some examples are

    "A string"
    "We can include any printing characters (e.g., &%@^*) in a string!!"
    "00001111ZZZZ"
    ""   – – *empty string*

If we need to include a double quotation mark character in a string, we write two double quotation mark characters together. The pair is interpreted as just one character in the string. For example:

    "A string in a string: ""A string"". "

If we need to write a string that is longer than will fit on one line, we can use the concatenation operator (&) to join two substrings together. (This operator is discussed in Chapter 4.) For example:

    "If a string will not fit on one line, "
    & "then we can break it into parts on separate lines."

Bit Strings

VHDL includes values that represent bits (binary digits), which can be either '0' or '1'. A bit-string literal represents a sequence of these bit values. It is represented by a string of digits, enclosed by double quotation marks and preceded by a character that specifies the base of the digits. The base specifier can be one of the following:

- B for binary,
- O for octal (base 8), and
- X for hexadecimal (base 16).

For example, some bit-string literals specified in binary are

> B"0100011"   B"10"   b"1111_0010_0001"   B""

Notice that we can include underline characters in bit-string literals to separate adjacent digits. The underline characters do not affect the meaning of the literal; they simply make the literal more readable. The base specifier can be in uppercase or lowercase. The last of the examples above denotes an empty bit string.

If the base specifier is octal, the digits '0' through '7' can be used. Each digit represents exactly three bits in the sequence. Some examples are

```
O"372"        – – equivalent to B"011_111_010"
o"00"         – – equivalent to B"000_000"
```

If the base specifier is hexadecimal, the digits '0' through '9' and 'A' through 'F' or 'a' through 'f' (representing 10 through 15) can be used. In hexadecimal, each digit represents exactly four bits. Some examples are

```
X"FA"         – – equivalent to B"1111_1010"
x"0d"         – – equivalent to B"0000_1101"
```

Notice that O"372" is not the same as X"FA", since the former is a sequence of nine bits, whereas the latter is a sequence of eight bits.

Syntax Descriptions

In the remainder of this book, we describe rules of syntax using a notation based on the Extended Backus-Naur Form (EBNF). These rules govern how we may combine lexical elements to form valid VHDL descriptions. It is useful to have a good working knowledge of the syntax rules, since VHDL analyzers expect valid VHDL descriptions as input. The error messages they otherwise produce may in some cases appear cryptic if we are unaware of the syntax rules.

The idea behind EBNF is to divide the language into *syntactic categories*. For each syntactic category we write a rule that describes how to build a VHDL clause of that category by combining lexical elements and clauses of other categories. These rules are analogous to the rules of English grammar. For example, there are rules that describe a sentence in terms of a subject and a predicate, and that describe a predicate in terms of a verb and an object phrase. In the rules for English grammar, "sentence", "subject", "predicate", and so on, are the syntactic categories.

In EBNF, we write a rule with the syntactic category we are defining on the left of a "⇐" sign (read as "is defined to be"), and a pattern on the right. The simplest kind of pattern is a collection of items in sequence, for example:

variable_assignment ⇐ target := expression ;

This rule indicates that a VHDL clause in the category "variable_assignment" is defined to be a clause in the category "target", followed by the symbol ":=", followed by a clause in the category "expression", followed by the symbol ";". To find out whether the VHDL clause

d0 := 25 + 6;

is syntactically valid, we would have to check the rules for "target" and "expression". As it happens, "d0" and "25 + 6" are valid subclauses, so the whole clause conforms to the pattern in the rule and is thus a valid variable assignment. On the other hand, the clause

25 fred := x **if** := .

cannot possibly be a valid variable assignment, since it doesn't match the pattern on the right side of the rule.

The next kind of rule to consider is one that allows for an optional component in a clause. We indicate the optional part by enclosing it between the symbols "⟦" and "⟧". For example:

function_call ⇐ name ⟦ (association_list) ⟧

This indicates that a function call consists of a name that may be followed by an association list in parentheses. Note the use of the outline symbols for writing the pattern in the rule, as opposed to the normal solid symbols that are lexical elements of VHDL.

In many rules, we need to specify that a clause is optional, but if present, it may be repeated as many times as needed. For example, in this simplified rule for a process statement:

process_statement ⇐
    process is
        { process_declarative_item }
    begin
        { sequential_statement }
    end process ;

the curly braces specify that a process may include zero or more process declarative items and zero or more sequential statements. A case that arises frequently in the rules of VHDL is a pattern consisting of some category followed by zero or more repetitions of that category. In this case, we use dots within the braces to represent the repeated category, rather than writing it out again in full. For example, the rule

```
case_statement ⇐
    case expression is
        case_statement_alternative
        { ₒₒₒ }
    end case ;
```

indicates that a case statement must contain at least one case statement alternative, but may contain an arbitrary number of additional case statement alternatives as required. If there is a sequence of categories and symbols preceding the braces, the dots represent only the last element of the sequence. Thus, in the example above, the dots represent only the case statement alternative, not the sequence "**case** expression **is** case_statement_alternative".

We also use the dots notation where a list of one or more repetitions of a clause is required, but some delimiter symbol is needed between repetitions. For example, the rule

```
identifier_list ⇐ identifier { , ₒₒₒ }
```

specifies that an identifier list consists of one or more identifiers, and that if there is more than one, they are separated by comma symbols. Note that the dots always represent a repetition of the category immediately preceding the left brace symbol. Thus, in the above rule, it is the identifier that is repeated, not the comma.

Many syntax rules allow a category to be composed of one of a number of alternatives. One way to represent this is to have a number of separate rules for the category, one for each alternative. However, it is often more convenient to combine alternatives using the "|" symbol. For example, the rule

```
mode ⇐ in | out | inout
```

specifies that the category "mode" can be formed from a clause consisting of one of the reserved words chosen from the alternatives listed.

The final notation we use in our syntax rules is parenthetic grouping, using the symbols "(" and ")". These simply serve to group part of a pattern, so that we can avoid any ambiguity that might otherwise arise. For example, the inclusion of parentheses in the rule

```
term ⇐ factor { ( * | / | mod | rem ) factor }
```

makes it clear that a factor may be followed by one of the operator symbols, and then another factor. Without the parentheses, the rule would be

```
term ⇐ factor { * | / | mod | rem factor }
```

indicating that a factor may be followed by one of the operators *, / or **mod** alone, or by the operator **rem** and then another factor. This is certainly not what is intended. The reason for this incorrect interpretation is that there is a *precedence*, or order of priority, in the EBNF notation we are using. In the absence of parentheses, a sequence of pattern components following one after the other is considered as a group with higher precedence than components separated by "|" symbols.

This EBNF notation is sufficient to describe the complete grammar of VHDL. However, there are often further constraints on a VHDL description that relate to the meaning of the lexical elements used. For example, a description specifying connection of a signal to a named object that identifies a component instead of a port is incorrect, even though it may conform to the syntax rules. To avoid such problems, many rules include additional information relating to the meaning of a language feature. For example, the rule shown above describing how a function call is formed is augmented thus:

function_call ⟸ *function*_name ⟦ (*parameter*_association_list) ⟧

The italicized prefix on a syntactic category in the pattern simply provides semantic information. This rule indicates that the name cannot be just any name, but must be the name of a function. Similarly, the association list must describe the parameters supplied to the function. (We will describe the meaning of functions and parameters in a later chapter.) The semantic information is for our benefit as designers reading the rule, to help us understand the intended semantics. So far as the syntax is concerned, the rule is equivalent to the original rule without the italicized parts.

In the following chapters, we will introduce each new feature of VHDL by describing its syntax using EBNF rules, and then we will describe the meaning and use of the feature through examples. In many cases, we will start with a simplified version of the syntax to make the description easier to learn and come back to the full details in a later chapter. For reference, Appendix C contains a complete listing of VHDL syntax in EBNF notation.

Exercises

1. [❶ 1.4] Briefly outline the purposes of the following VHDL modeling constructs: entity declaration, behavioral architecture body, structural architecture body, process statement, signal assignment statement and port map.

2. [❶ 1.5] Comment symbols are often used to make lines of a model temporarily ineffective. The symbol is added at the front of the line, turning the line into a comment. The comment symbol can be simply removed to reactivate the statement. The following process statement includes a line to assign a value to a test signal, to help debug the model. Modify the process to make the assignment ineffective.

   ```
   apply_transform : process is
   begin
       d_out <= transform(d_in) after 200 ps;
       debug_test <= transform(d_in);
       wait on enable, d_in;
   end process apply_transform;
   ```

3. [❶ 1.5] Which of the following are valid VHDL basic identifiers? Which are reserved words? Of the invalid identifiers, why are they invalid?

last_item	prev item	value–1	buffer
element#5	_control	93_999	entry_

4. [❶ 1.5] Rewrite the following decimal literals as hexadecimal literals.

1	34	256.0	0.5

5. [❶ 1.5] What decimal numbers are represented by the following literals?

     8#14#   2#1000_0100#   16#2C#

     2.5E5   2#1#E15   2#0.101#

6. [❶ 1.5] What is the difference between the literals 16#23DF# and X"23DF"?

7. [❶ 1.5] Express the following octal and hexadecimal bit strings as binary bit-string literals.

     O"747"   O"377"   O"1_345"

     X"F2"   X"0014"   X"0000_0001"

8. [❷ 1.4] Write an entity declaration and a behavioral architecture body for a two-input multiplexer, with input ports a, b and sel and an output port z. If the sel input is '0', the value of a should be copied to z, otherwise the value of b should be copied to z. Write a test bench for the multiplexer model, and test it using a VHDL simulator.

9. [❷ 1.4] Write an entity declaration and a structural architecture body for a four-bit-wide multiplexer, using instances of the two-bit multiplexer from Exercise 8. The input ports are a0, a1, a2, a3, b0, b1, b2, b3 and sel, and the output ports are z0, z1, z2 and z3. When sel is '0', the inputs a0 to a3 are copied to the outputs, otherwise the inputs b0 to b3 are copied to the outputs. Write a test bench for the multiplexer model, and test it using a VHDL simulator.

Scalar Data Types and Operations

The concept of type is very important when describing data in a VHDL model. The type of a data object defines the set of values that the object can assume, as well as the set of operations that can be performed on those values. A scalar type consists of single, indivisible values. In this chapter we look at the basic scalar types provided by VHDL and see how they can be used to define data objects that model the internal state of a module.

2.1   Constants and Variables

An *object* is a named item in a VHDL model that has a value of a specified type. There are four classes of objects: constants, variables, signals and files. In this chapter, we look at constants and variables; signals are described fully in Chapter 5. Files are considered an advanced topic and are not discussed in this book. Constants and variables are objects in which data can be stored for use in a model. The difference between them is that the value of a constant cannot be changed after it is created, whereas a variable's value can be changed as many times as necessary using variable assignment statements.

Constant and Variable Declarations

Both constants and variables need to be declared before they can be used in a model. A *declaration* simply introduces the name of the object, defines its type and may give it an initial value. The syntax rule for a constant declaration is

constant_declaration ⇐
    constant identifier { , ... } : subtype_indication ⟦ := expression ⟧ ;

The identifiers listed are the names of the constants being defined (one per name), and the subtype indication specifies the type of all of the constants. We look at ways of specifying the type in detail in subsequent sections of this chapter. The optional part shown in the syntax rule is an expression that specifies the value that each constant assumes. This part can only be omitted in certain cases that we discuss in Chapter 7. Until then, we always include it in examples. Here are some examples of constant declarations:

```
constant number_of_bytes : integer := 4;
constant number_of_bits : integer := 8 * number_of_bytes;
constant e : real := 2.718281828;
constant prop_delay : time := 3 ns;
constant size_limit, count_limit : integer := 255;
```

The reason for using a constant is to have a name and an explicitly defined type for a value, rather than just writing the value as a literal. This makes the model more intelligible to the reader, since the name and type convey much more information about the intended use of the object than the literal value alone. Furthermore, if we need to change the value as the model evolves, we only need to update the declaration. This is much easier and more reliable than trying to find and update all instances of a literal value throughout a model. It is good practice to use constants rather than writing literal values within a model.

The form of a variable declaration is similar to a constant declaration. The syntax rule is

variable_declaration ⇐
    variable identifier { , ... } : subtype_indication ⟦ := expression ⟧ ;

Here also the initialization expression is optional. If we omit it, the default initial value assumed by the variable when it is created depends on the type. For scalar types, the default initial value is the leftmost value of the type. For example, for integers it is the smallest representable integer. Some examples of variable declarations are

```
variable index : integer := 0;
variable sum, average, largest : real;
variable start, finish : time := 0 ns;
```

If we include more than one identifier in a variable declaration, it is the same as having separate declarations for each identifier. For example, the last declaration above is the same as the two declarations

```
variable start : time := 0 ns;
variable finish : time := 0 ns;
```

This is not normally significant unless the initialization expression is such that it potentially produces different values on two successive evaluations. The only time this may occur is if the initialization expression contains a call to a function with side effects (see Chapter 6).

Constant and variable declarations can appear in a number of places in a VHDL model, including in the declaration parts of processes. In this case, the declared object can be used only within the process. One restriction on where a variable declaration may occur is that it may not be placed so that the variable would be accessible to more than one process. This is to prevent the strange effects that might otherwise occur if the processes were to modify the variable in indeterminate order. The exception to this rule is if a variable is declared specially as a *shared* variable.

EXAMPLE

Figure 2-1 outlines an architecture body that shows how constant and variable declarations may be included in a VHDL model.

FIGURE 2-1

```
architecture sample of ent is
    constant pi : real := 3.14159;
begin
    process is
        variable counter : integer;
    begin
        . . .          -- statements using pi and counter
    end process;
end architecture sample;
```

An architecture body showing declarations of a constant pi *and a variable* counter.

Variable Assignment

Once a variable has been declared, its value can be modified by an assignment statement. The syntax of a variable assignment statement is given by the rule

variable_assignment_statement ⇐ ⟦ label : ⟧ name := expression ;

The optional label provides a means of identifying the assignment statement. The name in a variable assignment statement identifies the variable to be changed, and the expression is evaluated to produce the new value. The type of this value must match the type of the variable. The full details of how an expression is formed are covered in the rest of this chapter. For now, just think of expressions as the usual combinations of identifiers and literals with operators. Here are some examples of assignment statements:

```
program_counter := 0;
index := index + 1;
```

The first assignment sets the value of the variable **program_counter** to zero, overwriting any previous value. The second example increments the value of **index** by one.

It is important to note the difference between a variable assignment statement, shown here, and a signal assignment statement, introduced in Chapter 1. A variable assignment immediately overwrites the variable with a new value. A signal assignment, on the other hand, schedules a new value to be applied to a signal at some later time. We will return to signal assignments in Chapter 5. Because of the significant difference between the two kinds of assignment, VHDL uses distinct symbols: ":=" for variable assignment and "<=" for signal assignment.

VHDL-87

Variable assignment statements may not be labeled in VHDL-87.

2.2   Scalar Types

The notion of *type* is very important in VHDL. We say that VHDL is a *strongly typed* language, meaning that every object may only assume values of its nominated type. Furthermore, the definition of each operation includes the types of values to which the operation may be applied. The aim of strong typing is to allow detection of errors at an early stage of the design process, namely, when a model is analyzed.

In this section, we show how a new type is declared. We then show how to define different *scalar* types. A scalar type is one whose values are indivisible. In Chapter 4 we will show how to declare types whose values are composed of collections of element values.

Type Declarations

We introduce new types into a VHDL model by using type declarations. The declaration names a type and specifies which values may be stored in objects of the type. The syntax rule for a type declaration is

type_declaration ⇐ **type** identifier **is** type_definition ;

One important point to note is that if two types are declared separately with identical type definitions, they are nevertheless distinct and incompatible types. For example, if we have two type declarations:

```
type apples is range 0 to 100;
type oranges is range 0 to 100;
```

we may not assign a value of type **apples** to a variable of type **oranges**, since they are of different types.

An important use of types is to specify the allowed values for ports of an entity. In the examples in Chapter 1, we saw the type name **bit** used to specify that ports may take only the values '0' and '1'. If we define our own types for ports, the type names must be declared in a *package*, so that they are visible in the entity declaration. We will describe packages in more detail in Chapter 7; we introduce them here to enable us to write entity declarations using types of our own devising. For example, suppose we wish to define an adder entity that adds small integers in the range 0 to 255. We write a package containing the type declaration, as follows:

```
package int_types is
    type small_int is range 0 to 255;
end package int_types;
```

This defines a package named **int_types**, which provides the type named **small_int**. The package is a separate design unit and is analyzed before any entity declaration that needs to use the type it provides. We can use the type by preceding an entity declaration with a *use clause*, for example:

```
use work.int_types.all;
entity small_adder is
    port ( a, b : in small_int;  s : out small_int );
end entity small_adder;
```

When we discuss packages in Chapter 7, we will explain the precise meaning of use clauses such as this. For now, we treat it as "magic" needed to declare types for use in entity declarations.

Integer Types

In VHDL, integer types have values that are whole numbers. An example of an integer type is the predefined type **integer**, which includes all the whole numbers representable on a particular host computer. The language standard requires that the type **integer** include at least the numbers $-2{,}147{,}483{,}647$ to $+2{,}147{,}483{,}647$ ($-2^{31}+1$ to $+2^{31}-1$), but VHDL implementations may extend the range.

We can define a new integer type using a range-constraint type definition. The simplified syntax rule for an integer type definition is

integer_type_definition ⇐
    range simple_expression ⦅ **to** ‖ **downto** ⦆ simple_expression

which defines the set of integers between (and including) the values given by the two expressions. The expressions must evaluate to integer values. If we use the keyword **to**, we are defining an *ascending range*, in which values are ordered from the smallest on the left to the largest on the right. On the other hand, using the keyword **downto** defines a *descending range*, in which values are ordered left to right from largest to smallest. The reasons for distinguishing between ascending and descending ranges will become clear later.

EXAMPLE

Here are two integer type declarations:

type day_of_month **is range** 0 **to** 31;
type year **is range** 0 **to** 2100;

These two types are quite distinct, even though they include some values in common. Thus if we declare variables of these types:

variable today : day_of_month := 9;
variable start_year : year := 1987;

it would be illegal to make the assignment:

start_year := today;

Even though the number 9 is a member of the type **year**, in context it is treated as being of type **day_of_month**, which is incompatible with type **year**. This type rule helps us to avoid inadvertently mixing numbers that represent different kinds of things.

If we wish to use an arithmetic expression to specify the bounds of the range, the values used in the expression must be *locally static*, that is, they must be known when the model is analyzed. For example, we can use constant values in an expression as part of a range definition:

constant number_of_bits : integer := 32;
type bit_index **is range** 0 **to** number_of_bits − 1;

The operations that can be performed on values of integer types include the familiar arithmetic operations:

+	addition, or identity
−	subtraction, or negation
*	multiplication
/	division
mod	modulo
rem	remainder
abs	absolute value
**	exponentiation

The result of an operation is an integer of the same type as the operand or operands. For the binary operators (those that take two operands), the operands must be of the same type. The right operand of the exponentiation operator must be a non-negative integer.

The identity and negation operators are unary, meaning that they only take a single, right operand. The result of the identity operator is its operand unchanged, while the negation operator produces zero minus the operand. So, for example, the following all produce the same result:

A + (–B),   A – (+B),   A – B

The division operator produces an integer that is the result of dividing, with any fractional part truncated towards zero. The remainder operator is defined such that the relation

A = (A / B) * B + (A **rem** B)

is satisfied. The result of **A rem B** is the remainder left over from division of **A** by **B**. It has the same sign as **A** and has absolute value less than the absolute value of **B**. For example:

5 **rem** 3 = 2,   (–5) **rem** 3 = –2,   5 **rem** (–3) = 2,   (–5) **rem** (–3) = –2

Note that in these expressions, the parentheses are required by the grammar of VHDL. The two operators, **rem** and negation, may not be written side by side. The modulo operator conforms to the mathematical definition satisfying the relation

A = B * N + (A **mod** B)   – – for some integer N

The result of **A mod B** has the same sign as **B** and has absolute value less than the absolute value of **B**. For example:

5 **mod** 3 = 2,   (–5) **mod** 3 = 1,   5 **mod** (–3) = –1,   (–5) **mod** (–3) = –2

When a variable is declared to be of an integer type, the default initial value is the leftmost value in the range of the type. For ascending ranges, this will be the least value, and for descending ranges, it will be the greatest value. If we have these declarations:

```
type set_index_range is range 21 downto 11;
type mode_pos_range is range 5 to 7;
variable set_index : set_index_range;
variable mode_pos : mode_pos_range;
```

the initial value of **set_index** is 21, and that of **mode_pos** is 5. The initial value of a variable of type **integer** is –2,147,483,647 or less, since this type is predefined as an ascending range that must include –2,147,483,647.

Floating-Point Types

Floating-point types in VHDL are used to represent real numbers. Mathematically speaking, there is an infinite number of real numbers within any interval, so it is not possible to represent real numbers exactly on a computer. Hence floating-point types are only an approximation to real numbers. The term "floating point" refers to the fact that they are represented using a mantissa part and an exponent part. This is similar to the way in which we represent numbers in scientific notation.

There is a predefined floating-point type called **real**, which includes at least the range –1.0E+38 to +1.0E+38, with at least six decimal digits of precision. This corresponds to the IEEE standard 32-bit representation commonly used for floating-point numbers. Some implementations of VHDL may extend this range and may provide higher precision.

We define a new floating-point type using a range-constraint type definition. The simplified syntax rule for a floating-point type definition is

floating_type_definition ~
    range simple_expression (**to** ‖ **downto**) simple_expression

This is similar to the way in which an integer type is declared, except that the bounds must evaluate to floating-point numbers. Some examples of floating-point type declarations are

    type input_level **is range** –10.0 **to** +10.0;
    type probability **is range** 0.0 **to** 1.0;

The operations that can be performed on floating-point values include the arithmetic operations addition and identity (+), subtraction and negation (–), multiplication (*), division (/), absolute value (**abs**) and exponentiation (******). The result of an operation is of the same floating-point type as the operand or operands. For the binary operators (those that take two operands), the operands must be of the same type. The exception is that the right operand of the exponentiation operator must be an integer. The identity and negation operators are unary (meaning that they only take a single, right operand).

Variables that are declared to be of a floating-point type have a default initial value that is the leftmost value in the range of the type. So if we declare a variable to be of the type input_level shown above:

    variable input_A : input_level;

its initial value is –10.0.

Physical Types

The remaining numeric types in VHDL are physical types. They are used to represent real-world physical quantities, such as length, mass, time and current. The definition of a physical type includes the *primary unit* of measure and may also include some *secondary units*, which are integral multiples of the primary unit. The simplified syntax rule for a physical type definition is

physical_type_definition ~
    range simple_expression (**to** ‖ **downto**) simple_expression
      units
        identifier ;
        { identifier = physical_literal ; }
      end units [identifier]

This is like an integer type definition, but with the units definition part added. The primary unit (the first identifier after the **units** keyword) is the smallest unit that is represented. We may then define a number of secondary units, as we shall see in a moment. The range specifies the multiples of the primary unit that are included in the type. If the identifier is included at the end of the units definition part, it must repeat the name of the type being defined.

EXAMPLE

Here is a declaration of a physical type representing electrical resistance:

type resistance **is range** 0 **to** 1E9
    units
        ohm;
    end units resistance;

Literal values of this type are written as a numeric literal followed by the unit name, for example:

5 ohm   22 ohm   471_000 ohm

Notice that we must include a space before the unit name. Also, if the number is the literal 1, it can be omitted, leaving just the unit name. So the following two literals represent the same value:

ohm   1 ohm

Note that values such as –5 ohm and 1E16 ohm are not included in the type **resistance**, since the values –5 and **1E16** lie outside of the range of the type.

Now that we have seen how to write physical literals, we can look at how to specify secondary units in a physical type declaration. We do this by indicating how many primary units comprise a secondary unit. Our declaration for the resistance type can now be extended:

type resistance **is range** 0 **to** 1E9
    units
        ohm;
        kohm = 1000 ohm;
        Mohm = 1000 kohm;
    end units resistance;

Notice that once one secondary unit is defined, it can be used to specify further secondary units. Of course, the secondary units do not have to be powers of 10 times the primary unit; however, the multiplier must be an integer. For example, a physical type for length might be declared as

type length **is range** 0 **to** 1E9
    units
        um;   *– – primary unit: micron*
        mm = 1000 um;   *– – metric units*
        m = 1000 mm;
        mil = 254 um;   *– – imperial units*
        inch = 1000 mil;
    end units length;

We can write physical literals of this type using the secondary units, for example:

23 mm   450 mil   9 inch

When we write physical literals, we can write non-integral multiples of primary or secondary units. If the value we write is not an exact multiple of the primary unit, it is rounded down to the nearest multiple. For example, we might write the following literals of type **length**, each of which represents the same value:

0.1 inch   2.54 mm   2.540528 mm

The last of these is rounded down to **2540 um**, since the primary unit for **length** is **um**. If we write the physical literal **6.8 um**, it is rounded down to the value **6 um**.

Many of the arithmetic operators can be applied to physical types, but with some restrictions. The addition, subtraction, identity and negation operators can be applied to values of physical types, in which case they yield results that are of the same type as the operand or operands. A value of a physical type can be multiplied by an integer or a floating-point number to yield a value of the same physical type, for example:

5 mm * 6 = 30 mm

A value of a physical type can be divided by an integer or floating-point number to yield a value of the same physical type. Furthermore, two values of the same physical type can be divided to yield an integer, for example:

18 kohm / 2.0 = 9 kohm,   33 kohm / 22 ohm = 1500

Finally, the **abs** operator may be applied to a value of a physical type to yield a value of the same type, for example:

abs 2 foot = 2 foot,   **abs** (–2 foot) = 2 foot

The restrictions make sense when we consider that physical types represent actual physical quantities, and arithmetic should be done so as to produce results of the correct dimensions. It doesn't make sense to multiply two lengths together to yield a length; the result should logically be an area. So VHDL does not allow direct multiplication of two physical types. Instead, we must convert the values to abstract integers to do the calculation, then convert the result back to the final physical type. (See the discussion of the 'pos and 'val attributes in Section 2.4.)

A variable that is declared to be of a physical type has a default initial value that is the leftmost value in the range of the type. For example, the default initial values for the types declared above are **0 ohm** for **resistance** and **0 um** for **length**.

VHDL-87

A physical type definition in VHDL-87 may not repeat the type name after the keywords **end units**.

Time

The predefined physical type **time** is very important in VHDL, as it is used extensively to specify delays. Its definition is

```
type time is range implementation defined
    units
        fs;
        ps = 1000 fs;
        ns = 1000 ps;
        us = 1000 ns;
        ms = 1000 us;
        sec = 1000 ms;
        min = 60 sec;
        hr = 60 min;
    end units;
```

By default, the primary unit **fs** is the *resolution limit* used when a model is simulated. Time values smaller than the resolution limit are rounded down to zero units. A simulator may allow us to select a secondary unit of **time** as the resolution limit. In this case, the unit of all physical literals of type **time** in the model must not be less than the resolution limit. When the model is executed, the resolution limit is used to determine the precision with which time values are represented. The reason for allowing reduced precision in this way is to allow a greater range of time values to be represented. This may allow a model to be simulated for a longer period of simulation time.

Enumeration Types

Often when writing models of hardware at an abstract level, it is useful to use a set of names for the encoded values of some signals, rather than committing to a bit-level encoding straightaway. VHDL *enumeration types* allow us to do this. For example, suppose we are modeling a processor, and we want to define names for the function codes for the arithmetic unit. A suitable type declaration is

type alu_function **is** (disable, pass, add, subtract, multiply, divide);

Such a type is called an *enumeration*, because the literal values used are enumerated in a list. The syntax rule for enumeration type definitions in general is

enumeration_type_definition ⇐
    ((identifier ‖ character_literal) { , ∘∘∘ })

There must be at least one value in the type, and each value may be either an identifier, as in the above example, or a character literal. An example of this latter case is

type octal_digit **is** ('0', '1', '2', '3', '4', '5', '6', '7');

Given the above two type declarations, we could declare variables:

variable alu_op : alu_function;
variable last_digit : octal_digit := '0';

and make assignments to them:

alu_op := subtract;
last_digit := '7';

Different enumeration types may include the same identifier as a literal (called *over-loading*), so the context of use must make it clear which type is meant. To illustrate this, consider the following declarations:

```
type logic_level is (unknown, low, undriven, high);
variable control : logic_level;
type water_level is (dangerously_low, low, ok);
variable water_sensor : water_level;
```

Here, the literal **low** is overloaded, since it is a member of both types. However, the assignments

```
control := low;
water_sensor := low;
```

are both acceptable, since the types of the variables are sufficient to determine which **low** is being referred to.

When a variable of an enumeration type is declared, the default initial value is the leftmost element in the enumeration list. So **unknown** is the default initial value for type logic_level, and **dangerously_low** is that for type water_level.

There are three predefined enumeration types defined as

```
type severity_level is (note, warning, error, failure);
```

```
type file_open_status is (open_ok, status_error, name_error, mode_error);
```

```
type file_open_kind is (read_mode, write_mode, append_mode);
```

The type **severity_level** is used in assertion statements, which we will discuss in Chapter 3, and the types **file_open_status** and **file_open_kind** are used for file operations. (File operations are considered an advanced topic and not described in this book.) For the remainder of this section, we look at the other predefined enumeration types and the operations applicable to them.

VHDL-87

The types **file_open_status** and **file_open_kind** are not predefined in VHDL-87.

Characters

In Chapter 1 we saw how to write literal character values. These values are members of the predefined enumeration type **character**, which includes all of the characters in the ISO eight-bit character set. The type definition is shown in Figure 2-2. Note that this type is an example of an enumeration type containing a mixture of identifiers and character literals as elements.

The first 128 characters in this enumeration are the ASCII characters, which form a subset of the ISO character set. The identifiers from **nul** to **usp** and **del** are the non-printable ASCII control characters. Characters **c128** to **c159** do not have any standard names, so VHDL just gives them nondescript names based on their position in the character set. The character at position 160 is a non-breaking space character, distinct from the ordinary space character, and the character at position 173 is a soft hyphen.

To illustrate the use of the **character** type, we declare variables as follows:

FIGURE 2-2

type character **is** (

nul,	soh,	stx,	etx,	eot,	enq,	ack,	bel,
bs,	ht,	lf,	vt,	ff,	cr,	so,	si,
dle,	dc1,	dc2,	dc3,	dc4,	nak,	syn,	etb,
can,	em,	sub,	esc,	fsp,	gsp,	rsp,	usp,
' ',	'!',	'"',	'#',	'$',	'%',	'&',	''',
'(',	')',	'*',	'+',	',',	'-',	'.',	'/',
'0',	'1',	'2',	'3',	'4',	'5',	'6',	'7',
'8',	'9',	':',	';',	'<',	'=',	'>',	'?',
'@',	'A',	'B',	'C',	'D',	'E',	'F',	'G',
'H',	'I',	'J',	'K',	'L',	'M',	'N',	'O',
'P',	'Q',	'R',	'S',	'T',	'U',	'V',	'W',
'X',	'Y',	'Z',	'[',	'\',	']',	'^',	'_',
'`',	'a',	'b',	'c',	'd',	'e',	'f',	'g',
'h',	'i',	'j',	'k',	'l',	'm',	'n',	'o',
'p',	'q',	'r',	's',	't',	'u',	'v',	'w',
'x',	'y',	'z',	'{',	'|',	'}',	'~',	del
c128,	c129,	c130,	c131,	c132,	c133,	c134,	c135,
c136,	c137,	c138,	c139,	c140,	c141,	c142,	c143,
c144,	c145,	c146,	c147,	c148,	c149,	c150,	c151,
c152,	c153,	c154,	c155,	c156,	c157,	c158,	c159,
' ',	'¡',	'¢',	'£',	'¤',	'¥',	'¦',	'§'
'¨',	'©',	'ª',	'«',	'¬',	'-',	'®',	'¯'
'°',	'±',	'²',	'³',	'´',	'µ',	'¶',	'·',
'¸',	'¹',	'º',	'»',	'¼',	'½',	'¾',	'¿',
'À',	'Á',	'Â',	'Ã',	'Ä',	'Å',	'Æ',	'Ç',
'È',	'É',	'Ê',	'Ë',	'Ì',	'Í',	'Î',	'Ï',
'Đ',	'Ñ',	'Ò',	'Ó',	'Ô',	'Õ',	'Ö',	'×',
'Ø',	'Ù',	'Ú',	'Û',	'Ü',	'Ý',	'Þ',	'ß',
'à',	'á',	'â',	'ã',	'ä',	'å',	'æ',	'ç',
'è',	'é',	'ê',	'ë',	'ì',	'í',	'î',	'ï',
'ð',	'ñ',	'ò',	'ó',	'ô',	'õ',	'ö',	'÷',
'ø',	'ù',	'ú',	'û',	'ü',	'ý',	'þ',	'ÿ');

The definition of the predefined enumeration type character.

variable cmd_char, terminator : character;

and then make the assignments

cmd_char := 'P';
terminator := cr;

VHDL-87

Since VHDL-87 uses the ASCII character set, the predefined type **character** includes only the first 128 characters shown in Figure 2-2.

Booleans

One of the most important predefined enumeration types in VHDL is the type **boolean**, defined as

type boolean **is** (false, true);

This type is used to represent condition values, which can control execution of a behavioral model. There are a number of operators that we can apply to values of different types to yield Boolean values, namely, the relational and logical operators. The relational operators equality ("=") and inequality ("/=") can be applied to operands of any type (except files), including the composite types that we will see later in this chapter. The operands must both be of the same type, and the result is a Boolean value. For example, the expressions

123 = 123,   'A' = 'A',   7 ns = 7 ns

all yield the value **true**, and the expressions

123 = 456,   'A' = 'z',   7 ns = 2 us

yield the value **false**.

The relational operators that test ordering are the less-than ("<"), less-than-or-equal-to ("<="), greater-than (">") and greater-than-or-equal-to (">=") operators. These can only be applied to values of types that are ordered, including all of the scalar types described in this chapter. As with the equality and inequality operators, the operands must be of the same type, and the result is a Boolean value. For example, the expressions

123 < 456,   789 ps <= 789 ps,   '1' > '0'

all result in **true**, and the expressions

96 >= 102,   2 us < 4 ns,   'X' < 'X'

all result in **false**.

The logical operators **and**, **or**, **nand**, **nor**, **xor**, **xnor** and **not** take operands that must be Boolean values, and they produce Boolean results. Figure 2-3 shows the results produced by the binary logical operators. The result of the **not** operator is **true** if the operand is **false**, and **false** if the operand is **true**.

FIGURE 2-3

A	B	A **and** B	A **nand** B	A **or** B	A **nor** B	A **xor** B	A **xnor** B
false	false	false	true	false	true	false	true
false	true	false	true	true	false	true	false
true	false	false	true	true	false	true	false
true	true	true	false	true	false	false	true

The truth table for binary logical operators.

The operators **and**, **or**, **nand** and **nor** are called "short-circuit" operators, as they only evaluate the right operand if the left operand does not determine the result. For example, if the left operand of the **and** operator is false, we know that the result is false, so we do not need to consider the other operand. This is useful where the left operand is a test that guards against the right operand causing an error. Consider the expression

(b /= 0) **and** (a/b > 1)

If **b** were zero and we evaluated the right-hand operand, we would cause an error due to dividing by zero. However, because **and** is a short-circuit operator, if **b** were zero the left-hand operand would evaluate to false, so the right-hand operand would not be evaluated. For the **nand** operator, the right-hand operand is similarly not evaluated if the left-hand is false. For **or** and **nor**, the right-hand operand is not evaluated if the left-hand is true.

VHDL-87

The logical operator **xnor** is not provided in VHDL-87.

Bits

Since VHDL is used to model digital systems, it is useful to have a data type to represent bit values. The predefined enumeration type **bit** serves this purpose. It is defined as

type bit **is** ('0', '1');

Notice that the characters '0' and '1' are overloaded, since they are members of both **bit** and **character**. Where '0' or '1' occurs in a model, the context is used to determine which type is being used.

The logical operators that we mentioned for Boolean values can also be applied to values of type **bit**, and they produce results of type **bit**. The value '0' corresponds to false, and '1' corresponds to true. So, for example:

'0' **and** '1' = '0',   '1' **xor** '1' = '0'

The operands must still be of the same type as each other. Thus it is not legal to write

'0' **and** true

The difference between the types **boolean** and **bit** is that **boolean** values are used to model abstract conditions, whereas **bit** values are used to model hardware logic levels. Thus, '0' represents a low logic level and '1' represents a high logic level. The logical operators, when applied to **bit** values, are defined in terms of positive logic, with '0' representing the negated state and '1' representing the asserted state. If we need to deal with negative logic, we need to take care when writing logical expressions to get the correct logic sense. For example, if write_enable_n, select_reg_n and write_reg_n are negative logic bit variables, we perform the assignment

write_reg_n := **not** (**not** write_enable_n **and not** select_reg_n);

The variable write_reg_n is asserted ('0') only if write_enable_n is asserted and select_reg_n is asserted. Otherwise it is negated ('1').

Standard Logic

Since VHDL is designed for modeling digital hardware, it is necessary to include types to represent digitally encoded values. The predefined type **bit** shown above can be used for this in more abstract models, where we are not concerned about the details of electrical signals. However, as we refine our models to include more detail, we need to take account of the electrical properties when representing signals. There are many ways we can define data types to do this, but the IEEE has standardized one way in a package called **std_logic_1164**. The full details of the package are included in Appendix B. One of the types defined in this package is an enumeration type called **std_ulogic**, defined as

```
type std_ulogic is ( 'U',     – – Uninitialized
                     'X',     – – Forcing unknown
                     '0',     – – Forcing zero
                     '1',     – – Forcing one
                     'Z',     – – High impedance
                     'W',     – – Weak unknown
                     'L',     – – Weak zero
                     'H',     – – Weak one
                     '–' );   – – Don't care
```

This type can be used to represent signals driven by active drivers (forcing strength), resistive drivers such as pull-ups and pull-downs (weak strength) or three-state drivers including a high-impedance state. Each kind of driver may drive a "zero", "one" or "unknown" value. An "unknown" value is driven by a model when it is unable to determine whether the signal should be "zero" or "one". For example, the output of an and-gate is unknown when its inputs are driven by high-impedance drivers. In addition to these values, the leftmost value in the type represents an "uninitialized" value. If we declare signals of **std_ulogic** type, by default they take on 'U' as their initial value. If a model tries to operate on this value instead of a real logic value, we have detected a design error in that the system being modeled does not start up properly. The final value in **std_ulogic** is a "don't care" value. This is sometimes used by logic synthesis tools, and may also be used when defining test vectors, to denote that the value of a signal to be compared with a test vector is not important.

Even though the type **std_ulogic** and the other types defined in the **std_logic_1164** package are not actually built into the VHDL language, we can write models as though they were, with a little bit of preparation. For now, we describe some "magic" to include at the beginning of a model that uses the package; we explain the details in Chapter 7. If we include the line

library ieee; **use** ieee.std_logic_1164.**all**;

preceding each entity or architecture body that uses the package, we can write models as though the types were built into the language.

With this preparation in hand, we can now create constants, variables and signals of type **std_ulogic**. As well as assigning values of the type, we can also use the logical operators **and**, **or**, **not**, and so on. Each of these operates on **std_ulogic** values and returns a **std_ulogic** result of 'U', 'X', '0' or '1'. The operators are "optimistic," in that if they can determine a '0' or '1' result despite inputs being unknown, they do so. Otherwise they return 'X' or 'U'. For example **'0' and 'Z'** returns '0', since one input to an and-gate being '0' always causes the output to be '0', regardless of the other input.

2.3 Type Classification

In the preceding sections we have looked at the scalar types provided in VHDL. Figure 2-4 illustrates the relationships between these types, the predefined scalar types and the types we look at in later chapters.

FIGURE 2-4

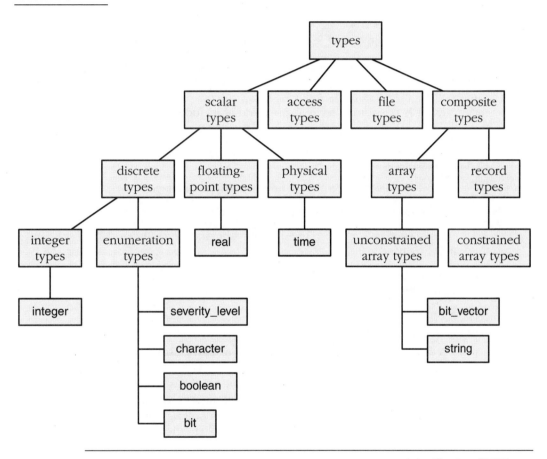

A classification of VHDL types.

The scalar types are all those composed of individual values that are ordered. Integer and floating-point types are ordered on the number line. Physical types are ordered by the number of base units in each value. Enumeration types are ordered by their declaration. The discrete types are those that represent discrete sets of values and comprise the integer types and enumeration types. Floating-point and physical types are not discrete, as they approximate a continuum of values.

Subtypes

In Section 2.2 we saw how to declare a type, which defines a set of values. Often a model contains objects that should only take on a restricted range of the complete set of values. We can represent such objects by declaring a *subtype*, which defines a restricted set of values from a *base type*. The condition that determines which values are in the subtype is called a *constraint*. Using a subtype declaration makes clear our intention about which values are valid and makes it possible to check that invalid values are not used. The simplified syntax rules for a subtype declaration are

subtype_declaration ⇐ **subtype** identifier **is** subtype_indication ;

subtype_indication ⇐
         name ⟦ **range** simple_expression (**to** ∥ **downto**) simple_expression ⟧

We will look at more advanced forms of subtype indications in later chapters. The subtype declaration defines the identifier as a subtype of the named base type, with the range constraint restricting the values for the subtype. The constraint is optional, which means that it is possible to have a subtype that includes all of the values of the base type.

EXAMPLE

Here is a declaration that defines a subtype of **integer**:

subtype small_int **is** integer **range** –128 **to** 127;

Values of **small_int** are constrained to be within the range –128 to 127. If we declare some variables:

variable deviation : small_int;
variable adjustment : integer;

we can use them in calculations:

deviation := deviation + adjustment;

Note that in this case, we can mix the subtype and base type values in the addition to produce a value of type **integer**, but the result must be within the range –128 to 127 for the assignment to succeed. If it is not, an error will be signaled when the variable is assigned. All of the operations that are applicable to the base type can also be used on values of a subtype. The operations produce values of the base type rather than the subtype. However, the assignment operation will not assign a value to a variable of a subtype if the value does not meet the constraint.

Another point to note is that if a base type is a range of one direction (ascending or descending), and a subtype is specified with a range constraint of the opposite direction, it is the subtype specification that counts. For example, the predefined type **integer** is an ascending range. If we declare a subtype as

subtype bit_index **is** integer **range** 31 **downto** 0;

this subtype is a descending range.

The VHDL standard includes two predefined integer subtypes, defined as

> **subtype** natural **is** integer **range** 0 **to** *highest integer*;
> **subtype** positive **is** integer **range** 1 **to** *highest integer*;

Where the logic of a design indicates that a number should not be negative, it is good style to use one of these subtypes rather than the base type **integer**. In this way, we can detect any design errors that incorrectly cause negative numbers to be produced. There is also a predefined subtype of the physical type **time**, defined as

> **subtype** delay_length **is** time **range** 0 fs **to** *highest time*;

This subtype should be used wherever a non-negative time delay is required.

VHDL-87

The subtype delay_length is not predefined in VHDL-87.

Type Qualification

Sometimes it is not clear from the context what the type of a particular value is. In the case of overloaded enumeration literals, it may be necessary to specify explicitly which type is meant. We can do this using *type qualification*, which consists of writing the type name followed by a single quote character, then an expression enclosed in parentheses. For example, given the enumeration types

> **type** logic_level **is** (unknown, low, undriven, high);
> **type** system_state **is** (unknown, ready, busy);

we can distinguish between the common literal values by writing

> logic_level'(unknown),   system_state'(unknown)

Type qualification can also be used to narrow a value down to a particular subtype of a base type. For example, if we define a subtype of logic_level

> **subtype** valid_level **is** logic_level **range** low **to** high;

we can explicitly specify a value of either the type or the subtype

> logic_level'(high),   valid_level'(high)

Of course, it is an error if the expression being qualified is not of the type or subtype specified.

Type Conversion

When we introduced the arithmetic operators in previous sections, we stated that the operands must be of the same type. This precludes mixing integer and floating-point values in arithmetic expressions. Where we need to do mixed arithmetic, we can use *type conversions* to convert between integer and floating-point values. The form of a type conversion is the name of the type we want to convert to, followed by a value in parentheses. For example, to convert between the types **integer** and **real**, we could write real(123),   integer(3.6)

Converting an integer to a floating-point value is simply a change in representation, although some loss of precision may occur. Converting from a floating-point value to an integer involves rounding to the nearest integer. Numeric type conversions are not the only conversion allowed. In general, we can convert between any closely related types. Other examples of closely related types are certain array types, discussed in Chapter 4.

One thing to watch out for is the distinction between type qualification and type conversion. The former simply states the type of a value, whereas the latter changes the value, possibly to a different type. One way to remember this distinction is to think of "*qu*ote for *qu*alification."

2.4   Attributes of Scalar Types

A type defines a set of values and a set of applicable operations. There is also a predefined set of *attributes* that are used to give information about the values included in the type. Attributes are written by following the type name with a quote mark (') and the attribute name. The value of an attribute can be used in calculations in a model. We now look at some of the attributes defined for the types we have discussed in this chapter.

First, there are a number of attributes that are applicable to all scalar types and provide information about the range of values in the type. If we let **T** stand for any scalar type or subtype, **x** stand for a value of that type and **s** stand for a string value, the attributes are

T'left	first (leftmost) value in T
T'right	last (rightmost) value in T
T'low	least value in T
T'high	greatest value in T
T'ascending	**true** if T is an ascending range, **false** otherwise
T'image(x)	a string representing the value of x
T'value(s)	the value in T that is represented by s

The string produced by the 'image attribute is a correctly formed literal according to the rules shown in Chapter 1. The strings allowed in the 'value attribute must follow those rules and may include leading or trailing spaces. These two attributes are useful for input and output in a model, as we will see when we come to that topic.

EXAMPLE

To illustrate the attributes listed above, we include some declarations from previous examples:

```
type resistance is range 0 to 1E9
    units
        ohm;
        kohm = 1000 ohm;
```

```
        Mohm = 1000 kohm;
    end units resistance;
```

type set_index_range **is range** 21 **downto** 11;

type logic_level **is** (unknown, low, undriven, high);

For these types:

```
resistance'left = 0 ohm
resistance'right = 1E9 ohm
resistance'low = 0 ohm
resistance'high = 1E9 ohm
resistance'ascending = true
resistance'image(2 kohm) = "2000 ohm"
resistance'value("5 Mohm") = 5_000_000 ohm
```

```
set_index_range'left = 21
set_index_range'right = 11
set_index_range'low = 11
set_index_range'high = 21
set_index_range'ascending = false
set_index_range'image(14) = "14"
set_index_range'value("20") = 20
```

```
logic_level'left = unknown
logic_level'right = high
logic_level'low = unknown
logic_level'high = high
logic_level'ascending = true
logic_level'image(undriven) = "undriven"
logic_level'value("Low") = low
```

Next, there are attributes that are applicable to just discrete and physical types. For any such type T, a value x of that type and an integer n, the attributes are

T'pos(x)	position number of x in T
T'val(n)	value in T at position n
T'succ(x)	value in T at position one greater than that of x
T'pred(x)	value in T at position one less than that of x
T'leftof(x)	value in T at position one to the left of x
T'rightof(x)	value in T at position one to the right of x

For enumeration types, the position numbers start at zero for the first element listed and increase by one for each element to the right. So, for the type logic_level shown above, some attribute values are

```
logic_level'pos(unknown) = 0
logic_level'val(3) = high
logic_level'succ(unknown) = low
logic_level'pred(undriven) = low
```

For integer types, the position number is the same as the integer value, but the type of the position number is a special anonymous type called *universal integer*. This is the same type as that of integer literals and, where necessary, is implicitly converted to any other declared integer type. For physical types, the position number is the integer number of base units in the physical value. For example:

time'pos(4 ns) = 4_000_000

since the base unit is fs.

EXAMPLE

We can use the 'pos and 'val attributes in combination to perform mixed-dimensional arithmetic with physical types, producing a result of the correct dimensionality. Suppose we define physical types to represent length and area, as follows:

```
type length is range integer'low to integer'high
    units
        mm;
    end units length;

type area is range integer'low to integer'high
    units
        square_mm;
    end units area;
```

and variables of these types:

```
variable L1, L2 : length;
variable A : area;
```

The restrictions on multiplying values of physical types prevents us from writing something like

```
A := L1 * L2;     – – this is incorrect
```

To achieve the correct result, we can convert the length values to abstract integers using the 'pos attribute, then convert the result of the multiplication to an area value using 'val, as follows:

```
A := area'val( length'pos(L1) * length'pos(L2) );
```

Note that in this example, we do not need to include a scale factor in the multiplication, since the base unit of **area** is the square of the base unit of **length**.

For ascending ranges, T'succ(x) and T'rightof(x) produce the same value, and T'pred(x) and T'leftof(x) produce the same value. For descending ranges, T'pred(x) and T'rightof(x) produce the same value, and T'succ(x) and T'leftof(x) produce the same value. For all ranges, T'succ(T'high), T'pred(T'low), T'rightof(T'right) and T'leftof(T'left) cause an error to occur.

The last attribute we introduce here is T'base. For any subtype T, this attribute produces the base type of T. The only context in which this attribute may be used is as the prefix of another attribute. For example, if we have the declarations

```
type opcode is (nop, load, store, add, subtract, negate, branch, halt);
subtype arith_op is opcode range add to negate;
```

then

```
arith_op'base'left = nop
arith_op'base'succ(negate) = branch
```

VHDL-87

The attributes 'ascending, 'image and 'value are not provided in VHDL-87.

2.5 Expressions and Operators

In Section 2.1 we showed how the value resulting from evaluation of an expression can be assigned to a variable. In this section, we summarize the rules governing expressions. We can think of an expression as being a formula that specifies how to compute a value. As such, it consists of primary values combined with operators. The precise syntax rules for writing expressions are shown in Appendix C. The primary values that can be used in expressions include

- literal values,
- identifiers representing data objects (constants, variables, and so on),
- attributes that yield values,
- qualified expressions,
- type-converted expressions, and
- expressions in parentheses.

We have seen examples of these in this chapter and in Chapter 1. For reference, all of the operators and the types they can be applied to are summarized in Figure 2-5. We will discuss array operators in Chapter 4.

The operators in this table are grouped by precedence, with ******, **abs** and **not** having highest precedence and the logical operators lowest. This means that if an expression contains a combination of operators, those with highest precedence are applied first. Parentheses can be used to alter the order of evaluation, or for clarity.

VHDL-87

The shift operators (**sll**, **srl**, **sla**, **sra**, **rol** and **ror**) and the **xnor** operator are not provided in VHDL-87.

FIGURE 2-5

Operator	Operation	Left operand type	Right operand type	Result type
**	exponentiation	integer or floating-point	integer	same as left operand
abs	absolute value		numeric	same as operand
not	negation		bit, boolean or 1-D array of bit or boolean	same as operand
*	multiplication	integer or floating-point	same as left operand	same as operands
		physical	integer or floating-point	same as left operand
		integer or floating-point	physical	same as right operand
/	division	integer or floating-point	same as left operand	same as operands
		physical	integer or floating-point	same as left operand
		physical	same as left operand	universal integer
mod	modulo	integer	same as left operand	same as operands
rem	remainder	integer	same as left operand	same as operands
+	identity		numeric	same as operand
−	negation		numeric	same as operand
+	addition	numeric	same as left operand	same as operands
-	subtraction	numeric	same as left operand	same as operands
&	concatenation	1-D array	same as left operand	same as operands
		1-D array	element type of left operand	same as left operand
		element type of right operand	1-D array	same as right operand
		element type of result	element type of result	1-D array
sll	shift-left logical	1-D array of bit or boolean	integer	same as left operand
srl	shift-right logical			
sla	shift-left arithmetic			
sra	shift-right arithmetic			
rol	rotate left			
ror	rotate right			

Operator	Operation	Left operand type	Right operand type	Result type
=	equality	any except file	same as left operand	**boolean**
/=	inequality			
<	less than	scalar or 1-D array	same as left operand	**boolean**
<=	less than or equal	of any discrete type		
>	greater than			
>=	greater than or equal			
and	logical and	**bit, boolean**	same as left operand	same as operands
or	logical or	or 1-D array of		
nand	negated logical and	**bit** or **boolean**		
nor	negated logical or			
xor	exclusive or			
xnor	negated exclusive or			

VHDL operators in order of precedence, from most-binding to least-binding.

Exercises

1. [❶ 2.1] Write constant declarations for the number of bits in a 32-bit word and for the number π (3.14159).

2. [❶ 2.1] Write variable declarations for a counter, initialized to 0; a status flag used to indicate whether a module is busy; and a standard-logic value used to store a temporary result.

3. [❶ 2.1] Given the declarations in Exercise 2, write variable assignment statements to increment the counter, to set the status flag to indicate the module is busy and to indicate a weak unknown temporary result.

4. [❶ 2.2] Write a package declaration containing type declarations for small non-negative integers representable in eight bits; fractional numbers between −1.0 and +1.0; electrical currents, with units of nA, μA, mA and A; and traffic light colors.

5. [❶ 2.4] Given the subtype declarations

   subtype pulse_range **is** time **range** 1 ms **to** 100 ms;
   subtype word_index **is** integer **range** 31 **downto** 0;

   what are the values of 'left, 'right, 'low, 'high and 'ascending attributes of each of these subtypes?

6. [❶ 2.4] Given the type declaration

    type state **is** (off, standby, active1, active2);

    what are the values of

    state'pos(standby)   state'val(2)
    state'succ(active2)   state'pred(active1)
    state'leftof(off)   state'rightof(off)

7. [❶ 2.5] For each of the following expressions, indicate whether they are syntactically correct, and if so, determine the resulting value.

    2 * 3 + 6 / 4   3 + –4
    "cat" & character'('0')   true **and** x **and not** y **or** z
    B"101110" **sll** 3   B"100010" **sra** 2 & X"2C"

8. [❷ 2.1] Write a counter model with a clock input **clk** of type **bit**, and an output **q** of type **integer**. The behavioral architecture body should contain a process that declares a count variable initialized to zero. The process should wait for changes on **clk**. When **clk** changes to '1', the process should increment the count and assign its value to the output port.

9. [❷ 2.2] Write a model that represents a simple ALU with integer inputs and output, and a function select input of type **bit**. If the function select is '0', the ALU output should be the sum of the inputs; otherwise the output should be the difference of the inputs.

10. [❷ 2.2] Write a model for a digital integrator that has a clock input of type **bit** and data input and output each of type **real**. The integrator maintains the sum of successive data input values. When the clock input changes from '0' to '1', the integrator should add the current data input to the sum and provide the new sum on the output.

11. [❷ 2.2] Following is a process that generates a regular clock signal.

    ```
    clock_gen : process is
    begin
        clk <= '1';  wait for 10 ns;
        clk <= '0';  wait for 10 ns;
    end process clock_gen;
    ```

    Use this as the basis for experiments to determine how your simulator behaves with different settings for the resolution limit. Try setting the resolution limit to 1 ns (the default for many simulators), 1 ps and 1 μs.

12. [❷ 2.2] Write a model for a tristate buffer using the standard-logic type for its data and enable inputs and its data output. If the enable input is '0' or 'L', the output should be 'Z'. If the enable input is '1' or 'H' and the data input is '0' or 'L', the output should be '0'. If the enable input is '1' or 'H' and the data input is '1' or 'H', the output should be '1'. In all other cases, the output should be 'X'.

Sequential Statements

In the previous chapter we saw how to represent the internal state of models using VHDL
data types. In this chapter we look at how that
data may be manipulated within processes.
This is done using *sequential statements*, so
called because they are executed in sequence.
We have already seen one of the basic sequential statements, the variable assignment statement, when we were looking at data types and
objects. The statements we look at in this chapter deal with controlling actions within a model,
hence they are often called *control structures*.
They allow selection between alternative
courses of action as well as repetition of actions.

3.1   If Statements

In many models, the behavior depends on a set of conditions that may or may not hold true during the course of simulation. We can use an *if statement* to express this behavior. The syntax rule for an if statement is

```
if_statement ⇐
    ⟦ if_label : ⟧
    if boolean_expression then
        { sequential_statement }
    { elsif boolean_expression then
        { sequential_statement } }
    ⟦ else
        { sequential_statement } ⟧
    end if ⟦ if_label ⟧ ;
```

At first sight, this may appear somewhat complicated, so we start with some simple examples and work up to examples showing the general case. The label may be used to identify the if statement. A simple example of an if statement is

```
if en = '1' then
    stored_value := data_in;
end if;
```

The expression after the keyword **if** is the condition that is used to control whether or not the statement after the keyword **then** is executed. If the condition evaluates to true, the statement is executed. In this example, if the value of the object **en** is '1', the assignment is made; otherwise it is skipped. We can also specify actions to be performed if the condition is false. For example:

```
if sel = 0 then
    result <= input_0;  – – executed if sel = 0
else
    result <= input_1;  – – executed if sel /= 0
end if;
```

Here, as the comments indicate, the first signal assignment statement is executed if the condition is true, and the second signal assignment statement is executed if the condition is false.

In many models, we may need to check a number of different conditions and execute a different sequence of statements for each case. We can construct a more elaborate form of if statement to do this, for example:

```
if mode = immediate then
    operand := immed_operand;
elsif opcode = load or opcode = add or opcode = subtract then
    operand := memory_operand;
else
    operand := address_operand;
end if;
```

In this example, the first condition is evaluated, and if true, the statement after the first **then** keyword is executed. If the first condition is false, the second condition is evaluated, and if it evaluates to true, the statement after the second **then** keyword is executed. If the second condition is false, the statement after the **else** keyword is executed.

In general, we can construct an if statement with any number of **elsif** clauses (including none), and we may include or omit the **else** clause. Execution of the if statement starts by evaluating the first condition. If it is false, successive conditions are evaluated, in order, until one is found to be true, in which case the corresponding statements are executed. If none of the conditions is true, and we have included an **else** clause, the statements after the **else** keyword are executed.

We are not restricted to just one statement in each part of the if statement. This is illustrated by the following if statement:

```
if opcode = halt_opcode then
    PC := effective_address;
    executing := false;
    halt_indicator <= true;
end if;
```

If the condition is true, all three statements are executed, one after another. On the other hand, if the condition is false, none of the statements are executed. Furthermore, each statement contained in an if statement can be any sequential statement. This means we can nest if statements, for example:

```
if phase = wash then
    if cycle_select = delicate_cycle then
        agitator_speed <= slow;
    else
        agitator_speed <= fast;
    end if;
    agitator_on <= true;
end if;
```

In this example, the condition **phase = wash** is first evaluated, and if true, the nested if statement and the following signal assignment statement are executed. Thus the assignment **agitator_speed <= slow** is executed only if both conditions evaluate to true, and the assignment **agitator_speed <= fast** is executed only if the first condition is true and the second condition is false.

EXAMPLE

Let us develop a behavioral model for a simple heater thermostat. The device can be modeled as an entity with two integer inputs, one that specifies the desired temperature and another that is connected to a thermometer, and one Boolean output that turns a heater on and off. The thermostat turns the heater on if the measured temperature falls below two degrees less than the desired temperature, and turns the heater off if the measured temperature rises above two degrees greater than the desired temperature. Figure 3-1 shows the entity and architecture bodies for the thermostat. The entity declaration defines the input and output ports.

FIGURE 3-1

```
entity thermostat is
    port ( desired_temp, actual_temp : in integer;
            heater_on : out boolean );
end entity thermostat;

- - - - - - - - - - - - - - - - - - - - - - - - - - - - - - - - - - - - -

architecture example of thermostat is
begin
    controller : process (desired_temp, actual_temp) is
    begin
        if actual_temp < desired_temp – 2 then
            heater_on <= true;
        elsif actual_temp > desired_temp + 2 then
            heater_on <= false;
        end if;
    end process controller;
end architecture example;
```

An entity and architecture body for a heater thermostat.

Since it is a behavioral model, the architecture body contains only a process statement that implements the required behavior. The process statement includes a *sensitivity list* after the keyword **process**. This is a list of signals to which the process is sensitive. When any of these signals change value, the process resumes and executes the sequential statements. After it has executed the last statement, the process suspends again. In this example, the process is sensitive to changes on either of the input ports. Thus, if we adjust the desired temperature, or if the measured temperature from the thermometer varies, the process is resumed. The body of the process contains an if statement that compares the actual temperature with the desired temperature. If the actual temperature is too low, the process executes the first signal assignment to turn the heater on. If the actual temperature is too high, the process executes the second signal assignment to turn the heater off. If the actual temperature is within the range, the state of the heater is not changed, since there is no **else** clause in the if statement.

VHDL-87

If statements may not be labeled in VHDL-87.

3.2   Case Statements

If we have a model in which the behavior is to depend on the value of a single expression, we can use a *case statement*. The syntax rules are as follows:

```
case_statement ⇐
    〚 case_label : 〛
    case expression is
        ( when choices => { sequential_statement } )
        { ... }
    end case 〚 case_label 〛 ;

choices ⇐ ( simple_expression ‖ discrete_range ‖ others ) { | ... }
```

The label may be used to identify the case statement. We start with some simple examples of case statements and build up from them. First, suppose we are modeling an arithmetic/logic unit, with a control input, func, declared to be of the enumeration type:

```
type alu_func is (pass1, pass2, add, subtract);
```

We could describe the behavior using a case statement:

```
case func is
    when pass1 =>
        result := operand1;
    when pass2 =>
        result := operand2;
    when add =>
        result := operand1 + operand2;
    when subtract =>
        result := operand1 – operand2;
end case;
```

At the head of this case statement is the *selector expression*, between the keywords **case** and **is**. In this example it is a simple expression consisting of just a primary value. The value of this expression is used to select which statements to execute. The body of the case statement consists of a series of *alternatives*, each starting with the keyword **when**, followed by one or more *choices* and a sequence of statements. The choices are values that are compared with the value of the selector expression. There must be exactly one choice for each possible value of the selector expression. The case statement finds the alternative with the same choice value as the selector expression and executes the statements in that alternative. In this example, the choices are all simple expressions of type alu_func. If the value of func is pass1, the statement result := operand1 is executed; if the value is pass2, the statement result := operand2 is executed; and so on.

A case statement bears some similarity to an if statement in that they both select among alternative groups of sequential statements. The difference lies in how the statements to be executed are chosen. We saw in the previous section that an if statement evaluates successive Boolean conditions in turn until one is found to be true. The group of statements corresponding to that condition is then executed. A case statement, on the other hand, evaluates a single selector expression to derive a selector value. This value is then compared with the choice values in the case statement alternatives to determine which statement to execute. An if statement provides a more general mechanism for selecting between alternatives, since the conditions can be arbitrarily complex Boolean expressions. However, case statements are an important and useful modeling mechanism, as the examples in this section show.

The selector expression of a case statement must result in a value of a discrete type, or a one-dimensional array of character elements, such as a character string or bit string (see Chapter 4). Thus, we can have a case statement that selects an alternative based on an integer value. If we assume index_mode and instruction_register are declared as

subtype index_mode **is** integer **range** 0 **to** 3;

variable instruction_register : integer **range** 0 **to** 2**16 – 1;

then we can write a case statement that uses a value of this type:

```
case index_mode'((instruction_register / 2**12) rem 2**2) is
    when 0 =>
        index_value := 0;
    when 1 =>
        index_value := accumulator_A;
    when 2 =>
        index_value := accumulator_B;
    when 3 =>
        index_value := index_register;
end case;
```

Notice that in this example, we use a qualified expression in the selector expression. If we had omitted this, the result of the expression would have been integer, and we would have had to include alternatives to cover all possible integer values. The type qualification avoids this need by limiting the possible values of the expression.

Another rule to remember is that the type of each choice must be the same as the type resulting from the selector expression. Thus in the above example, it is illegal to include an alternative such as

when 'a' => ...   *-- illegal!*

since the choice listed cannot be an integer. Such a choice does not make sense, since it can never match a value of type integer.

We can include more than one choice in each alternative by writing the choices separated by the "|" symbol. For example, if the type opcodes is declared as

type opcodes **is**
        (nop, add, subtract, load, store, jump, jumpsub, branch, halt);

we could write an alternative including three of these values as choices:

when load | add | subtract =>
        operand := memory_operand;

If we have a number of alternatives in a case statement and we want to include an alternative to handle all possible values of the selector expression not mentioned in previous alternatives, we can use the special choice **others**. For example, if the variable **opcode** is a variable of type **opcodes**, declared above, we can write

```
case opcode is
    when load I add I subtract =>
        operand := memory_operand;
    when store I jump I jumpsub I branch =>
        operand := address_operand;
    when others =>
        operand := 0;
end case;
```

In this example, if the value of **operand** is anything other than the choices listed in the first and second alternatives, the last alternative is selected. There may only be one alternative that uses the **others** choice, and if it is included, it must be the last alternative in the case statement. An alternative that includes the **others** choice may not include any other choices. Note that, if all of the possible values of the selector expression are covered by previous choices, we may still include the **others** choice, but it can never be matched.

The remaining form of choice that we have not yet mentioned is a *discrete range*, specified by these simplified syntax rules:

discrete_range ⇐
    *discrete*_subtype_indication
    ‖ simple_expression (**to** ‖ **downto**) simple_expression

subtype_indication ⇐
    type_mark
        ⟦ **range** simple_expression (**to** ‖ **downto**) simple_expression ⟧

These forms allow us to specify a range of values in a case statement alternative. If the value of the selector expression matches any of the values in the range, the statements in the alternative are executed. The simplest way to specify a discrete range is just to write the left and right bounds of the range, separated by a direction keyword. For example, the case statement above could be rewritten as

```
case opcode is
    when add to load =>
        operand := memory_operand;
    when branch downto store =>
        operand := address_operand;
    when others =>
        operand := 0;
end case;
```

Another way of specifying a discrete range is to use the name of a discrete type, and possibly a range constraint to narrow down the values to a subset of the type. For example, if we declare a subtype of **opcodes** as

subtype control_transfer_opcodes **is** opcodes **range** jump **to** branch;

we can rewrite the second alternative as

```
        when control_transfer_opcodes | store =>
            operand := address_operand;
```

Note that we may only use a discrete range as a choice if the selector expression is of a discrete type. We may not use a discrete range if the selector expression is of an array type, such as a bit-vector type. If we specify a range by writing the bounds and a direction, the direction has no significance except to identify the contents of the range.

An important point to note about the choices in a case statement is that they must all be written using *locally static* values. This means that the values of the choices must be determined during the analysis phase of design processing. All of the above examples satisfy this requirement. To give an example of a case statement that fails this requirement, suppose we have an integer variable N, declared as

```
    variable N : integer := 1;
```

If we wrote the case statement

```
    case expression is               -- example of an illegal case statement
        when N | N+1 => . . .
        when N+2 to N+5 => . . .
        when others => . . .
    end case;
```

the values of the choices depend on the value of the variable N. Since this might change during the course of execution, these choices are not locally static. Hence the case statement as written is illegal. On the other hand, if we had declared C to be a constant integer, for example with the declaration

```
    constant C : integer := 1;
```

then we could legally write the case statement

```
    case expression is
        when C | C+1 => . . .
        when C+2 to C+5 => . . .
        when others => . . .
    end case;
```

This is legal, since we can determine, by analyzing the model, that the first alternative includes choices 1 and 2, the second includes numbers between 3 and 6 and the third covers all other possible values of the expression.

The previous examples all show only one statement in each alternative. As with the if statement, we can write an arbitrary number of sequential statements of any kind in each alternative. This includes writing nested case statements, if statements or any other form of sequential statements in the alternatives.

Although the preceding rules governing case statements may seem complex, in practice there are just a few things to remember, namely:

- all possible values of the selector expression must be covered by one and only one choice,
- the values in the choices must be locally static, and

- if the **others** choice is used it must be in the last alternative and must be the only choice in that alternative.

EXAMPLE

We can write a behavioral model of a multiplexer with a select input sel; four data inputs d0, d1, d2 and d3; and a data output z. The data inputs and outputs are of the IEEE standard-logic type, and the select input is of type sel_range, which we assume to be declared elsewhere as

type sel_range **is range** 0 **to** 3;

We show in Chapter 7, when we discuss packages, how we define a type for use in an entity declaration. The entity declaration defining the ports and a behavioral architecture body are shown in Figure 3-2. The architecture body contains just a process declaration. Since the output of the multiplexer must change if any of the data or select inputs change, the process must be sensitive to all of the inputs. It makes use of a case statement to select which of the data inputs is to be assigned to the data output.

FIGURE 3-2

```
library ieee;  use ieee.std_logic_1164.all;
entity mux4 is
    port ( sel : in sel_range;
           d0, d1, d2, d3 : in std_ulogic;
           z : out std_ulogic );
end entity mux4;
– – – – – – – – – – – – – – – – – – – – – – – – – – – – – – – – – – – – –
architecture demo of mux4 is
begin
    out_select : process (sel, d0, d1, d2, d3) is
    begin
        case sel is
            when 0 =>
                z <= d0;
            when 1 =>
                z <= d1;
            when 2 =>
                z <= d2;
            when 3 =>
                z <= d3;
        end case;
    end process out_select;
end architecture demo;
```

An entity and architecture body for a four-input multiplexer.

VHDL-87

Case statements may not be labeled in VHDL-87.

3.3   Null Statements

Sometimes when writing models we need to state that when some condition arises, no action is to be performed. This need often arises when we use case statements, since we must include an alternative for every possible value of the selector expression. Rather than just leaving the statement part of an alternative blank, we can use a *null statement* to state explicitly that nothing is to be done. The syntax rule for the null statement is simply

> null_statement ⇐ ⟦ label : ⟧ **null** ;

The optional label serves to identify the statement. A simple, unlabeled null statement is

> **null**;

An example of its use in a case statement is

```
case opcode is
    when add =>
        Acc := Acc + operand;
    when subtract =>
        Acc := Acc – operand;
    when nop =>
        null;
end case;
```

We can use a null statement in any place where a sequential statement is required, not just in a case statement alternative. A null statement may be used during the development phase of model writing. If we know, for example, that we will need an entity as part of a system, but we are not yet in a position to write a detailed model for it, we can write a behavioral model that does nothing. Such a model just includes a process with a null statement in its body:

```
control_section : process ( sensitivity-list ) is
begin
    null;
end process control_section;
```

Note that the process must include the sensitivity list, for reasons that are explained in Chapter 5.

VHDL-87

Null statements may not be labeled in VHDL-87.

3.4   Loop Statements

Often we need to write a sequence of statements that is to be repeatedly executed. We use a *loop statement* to express this behavior. There are several different forms of loop statements in VHDL; the simplest is a loop that repeats a sequence of statements indefinitely, often called an *infinite loop*. The syntax rule for this kind of loop is

> loop_statement ⟸
>   [*loop*_label :]
>   **loop**
>   { sequential_statement }
>   **end loop** [*loop*_label] ;

In most computer programming languages, an infinite loop is not desirable, since it means that the program never terminates. However, when we are modeling digital systems, an infinite loop can be useful, since many hardware devices repeatedly perform the same function until we turn off the power. Typically a model for such a system includes a loop statement in a process body; the loop, in turn, contains a wait statement.

EXAMPLE

Figure 3-3 is a model for a counter that starts from zero and increments on each clock transition from '0' to '1'. When the counter reaches 15, it wraps back to zero on the next clock transition. The architecture body for the counter contains a process that first initializes the **count** output to zero, then repeatedly waits for a clock transition before incrementing the count value.

FIGURE 3-3

```
entity counter is
    port ( clk : in bit;  count : out natural );
end entity counter;
- - - - - - - - - - - - - - - - - - - - - - - - - - - - - - - - - - - - - -
architecture behavior of counter is
begin
    incrementer : process is
        variable count_value : natural := 0;
    begin
        count <= count_value;
        loop
            wait until clk = '1';
            count_value := (count_value + 1) mod 16;
            count <= count_value;
        end loop;
    end process incrementer;
end architecture behavior;
```

An entity and architecture body for a counter.

The wait statement in this example causes the process to suspend in the middle of the loop. When the clk signal changes from '0' to '1', the process resumes and updates the count value and the **count** output. The loop is then repeated starting with the wait statement, so the process suspends again.

Another point to note in passing is that the process statement does not include a sensitivity list. This is because it includes a wait statement. A process may contain either a sensitivity list or wait statements, but not both. We will return to this in detail in Chapter 5.

Exit Statements

In the previous example, the loop repeatedly executes the enclosed statements, with no way of stopping. Usually we need to exit the loop when some condition arises. We can use an *exit statement* to exit a loop. The syntax rule is

exit_statement ⟸
    〚 label : 〛 **exit** 〚 *loop*_label 〛 〚 **when** *boolean*_expression 〛 ;

The optional label at the start of the exit statement serves to identify the statement. The simplest form of exit statement is just

exit;

When this statement is executed, any remaining statements in the loop are skipped, and control is transferred to the statement after the **end loop** keywords. So in a loop we can write

if *condition* **then**
    exit;
end if;

where *condition* is a Boolean expression. Since this is perhaps the most common use of the exit statement, VHDL provides a shorthand way of writing it, using the **when** clause. We use an exit statement with the **when** clause in a loop of the form

loop
    . . .
    exit when *condition*;
    . . .
end loop;
    . . .   – – *control transferred to here*
                 – – *when* condition *becomes true within the loop*

EXAMPLE

We now revise the previous counter model to include a **reset** input that, when '1', causes the **count** output to be reset to zero. The output stays at zero as long as the **reset** input is '1' and resumes counting on the next clock transition after **reset** changes to '0'. The revised entity declaration, shown in Figure 3-4, includes the new input port.

FIGURE 3-4

```
entity counter is
    port ( clk, reset : in bit;  count : out natural );
end entity counter;
```

```
architecture behavior of counter is
begin
    incrementer : process is
        variable count_value : natural := 0;
    begin
        count <= count_value;
        loop
            loop
                wait until clk = '1' or reset = '1';
                exit when reset = '1';
                count_value := (count_value + 1) mod 16;
                count <= count_value;
            end loop;
            – – at this point, reset = '1'
            count_value := 0;
            count <= count_value;
            wait until reset = '0';
        end loop;
    end process incrementer;
end architecture behavior;
```

An entity and architecture body of the revised counter, including a reset *input.*

The architecture body is revised by nesting the loop inside another loop statement and adding the **reset** signal to the original wait statement. The inner loop
performs the same function as before, except that when **reset** changes to '1', the
process is resumed, and the exit statement causes the inner loop to be terminated.
Control is transferred to the statement just after the end of the inner loop. As the
comment indicates, we know that this point can only be reached when **reset** is '1'.
The count value and **count** outputs are reset, and the process then waits for **reset**
to return to '0'. While it is suspended at this point, any changes on the clock input
are ignored. When **reset** changes to '0', the process resumes, and the outer loop
repeats.

This example also illustrates another important point. When we have nested
loop statements, with an exit statement inside the inner loop, the exit statement
causes control to be transferred out of the inner loop only, not the outer loop.
By default, an exit statement transfers control out of the immediately enclosing
loop.

In some cases, we may wish to transfer control out of an inner loop and also a containing loop. We can do this by labeling the outer loop and using the label in the exit statement. We can write

```
loop_name : loop
    . . .
    exit loop_name;
    . . .
end loop loop_name ;
```

This labels the loop with the name loop_name, so that we can indicate which loop to exit in the exit statement. The loop label can be any valid identifier. The exit statement referring to this label can be located within nested loop statements.

To illustrate how loops can be nested, labeled and exited, let us consider the following statements:

```
outer : loop
    . . .
    inner : loop
        . . .
        exit outer when condition-1;   – – exit 1
        . . .
        exit when condition-2;          – – exit 2
        . . .
    end loop inner;
    . . .                               – – target A
    exit outer when condition-3;        – – exit 3
    . . .
end loop outer;
    . . .                               – – target B
```

This example contains two loop statements, one labeled inner nested inside another labeled outer. The first exit statement, tagged with the comment exit 1, transfers control to the statement tagged target B if its condition is true. The second exit statement, tagged exit 2, transfers control to target A. Since it does not refer to a label, it only exits the immediately enclosing loop statement, namely, loop inner. Finally, the exit statement tagged exit 3 transfers control to target B.

VHDL-87

Exit statements may not be labeled in VHDL-87.

Next Statements

Another kind of statement that we can use to control the execution of loops is the *next statement*. When this statement is executed, the current iteration of the loop is completed without executing any further statements, and the next iteration is begun. The syntax rule is

next_statement ⇐
    〖 label : 〗 **next** 〖 *loop*_label 〗 〖 **when** *boolean*_expression 〗 ;

The optional label at the start of the next statement serves to identify the statement. A next statement is very similar in form to an exit statement, the difference being the keyword **next** instead of **exit**. The simplest form of next statement is

    next;

which starts the next iteration of the immediately enclosing loop. We can also include a condition to test before completing the iteration:

    next when *condition*;

and we can include a loop label to indicate for which loop to complete the iteration:

    next *loop-label*;

or:

    next *loop-label* **when** *condition*;

A next statement that exits the immediately enclosing loop can be easily rewritten as an equivalent loop with an if statement replacing the next statement. For example, the following two loops are equivalent:

loop	**loop**
statement-1;	*statement-1*;
next when *condition*;	**if not** *condition* **then**
statement-2;	*statement-2*;
end loop;	**end if**;
	end loop;

However, nested labeled loops that contain next statements referring to outer loops cannot be so easily rewritten. As a matter of style, if we find ourselves about to write such a collection of loops and next statements, it's probably time to think more carefully about what we are trying to express. If we check the logic of the model, we may be able to find a simpler formulation of loop statements. Complicated loop/next structures can be confusing, making the model hard to read and understand.

VHDL-87

Next statements may not be labeled in VHDL-87.

While Loops

We can augment the basic loop statement introduced previously to form a *while loop*, which tests a condition before each iteration. If the condition is true, iteration proceeds. If it is false, the loop is terminated. The syntax rule for a while loop is

```
loop_statement ⇐
    ⟦ loop_label : ⟧
    while condition loop
        { sequential_statement }
    end loop ⟦ loop_label ⟧ ;
```

The only difference between this form and the basic loop statement is that we have added the keyword **while** and the condition before the **loop** keyword. All of the things we said about the basic loop statement also apply to a while loop. We can write any sequential statements in the body of the loop, including exit and next statements, and we can label the loop by writing the label before the **while** keyword.

There are three important points to note about while loops. The first point is that the condition is tested before each iteration of the loop, including the first iteration. This means that if the condition is false before we start the loop, it is terminated immediately, with no iterations being executed. For example, given the while loop

```
while index > 0 loop
    . . .              -- statement A: do something with index
end loop;
    . . .              -- statement B
```

if we can demonstrate that **index** is not greater than zero before the loop is started, then we know that the statements inside the loop will not be executed, and control will be transferred straight to **statement B**.

The second point is that in the absence of exit statements within a while loop, the loop terminates only when the condition becomes false. Thus, we know that the negation of the condition must hold when control reaches the statement after the loop. Similarly, in the absence of next statements within a while loop, the loop performs an iteration only when the condition is true. Thus, we know that the condition holds when we start the statements in the loop body. In the above example, we know that **index** must be greater then zero when we execute the statement tagged **statement A**, and also that **index** must be less than or equal to zero when we reach **statement B**. This knowledge can help us reason about the correctness of the model we are writing.

The third point is that when we write the statements inside the body of a while loop, we must make sure that the condition will eventually become false, or that an exit statement will eventually exit the loop. Otherwise the while loop will never terminate. Presumably, if we had intended to write an infinite loop, we would have used a simple loop statement.

EXAMPLE

We can develop a model for an entity cos that might be used as part of a specialized signal processing system. The entity has one input, theta, which is a real number representing an angle in radians, and one output, result, representing the cosine function of the value of theta. We can use the relation

$$\cos \theta = 1 - \frac{\theta^2}{2!} + \frac{\theta^4}{4!} - \frac{\theta^6}{6!} + \cdots$$

by adding successive terms of the series until the terms become smaller than one millionth of the result. The entity and architecture body declarations are shown in Figure 3-5.

FIGURE 3-5

```
entity cos is
    port ( theta : in real;  result : out real );
end entity cos;

- - - - - - - - - - - - - - - - - - - - - - - - - - - - - - - - - - - - - - - -

architecture series of cos is
begin
    summation : process (theta) is
        variable sum, term : real;
        variable n : natural;
    begin
        sum := 1.0;
        term := 1.0;
        n := 0;
        while abs term > abs (sum / 1.0E6) loop
            n := n + 2;
            term := (–term) * theta**2 / real(((n–1) * n));
            sum := sum + term;
        end loop;
        result <= sum;
    end process summation;
end architecture series;
```

An entity and architecture body for a cosine module.

The architecture body consists of a process that is sensitive to changes in the input signal theta. Initially, the variables sum and term are set to 1.0, representing the first term in the series. The variable n starts at 0 for the first term. The cosine function is computed using a while loop that increments n by two and uses it to calculate the next term based on the previous term. Iteration proceeds as long as the last term computed is larger in magnitude than one millionth of the sum. When the last term falls below this threshold, the while loop is terminated. We can determine that the loop will terminate, since the values of successive terms in the series get progressively smaller. This is because the factorial function grows at a greater rate than the exponential function.

For Loops

Another way we can augment the basic loop statement is the *for loop*. A for loop includes a specification of how many times the body of the loop is to be executed. The syntax rule for a for loop is

```
loop_statement ⇐
    ⟦ loop_label : ⟧
    for identifier in discrete_range loop
        { sequential_statement }
    end loop ⟦ loop_label ⟧ ;
```

We saw on page 59 that a discrete range can be of the form

simple_expression (**to** ‖ **downto**) simple_expression

representing all the values between the left and right bounds, inclusive. The identifier is called the *loop parameter,* and for each iteration of the loop, it takes on successive values of the discrete range, starting from the left element. For example, in this for loop:

```
for count_value in 0 to 127 loop
    count_out <= count_value;
    wait for 5 ns;
end loop;
```

the identifier count_value takes on the values 0, 1, 2 and so on, and for each value, the assignment and wait statements are executed. Thus the signal count_out will be assigned values 0, 1, 2 and so on, up to 127, at 5 ns intervals.

We also saw that a discrete range can be specified using a discrete type or subtype name, possibly further constrained to a subset of values by a range constraint. For example, if we have the enumeration type

```
type controller_state is (initial, idle, active, error);
```

we can write a for loop that iterates over each of the values in the type

```
for state in controller_state loop
    . . .
end loop;
```

Within the sequence of statements in the for loop body, the loop parameter is a constant whose type is the base type of the discrete range. This means we can use its value by including it in an expression, but we cannot make assignments to it. Unlike other constants, we do not need to declare it. Instead, the loop parameter is implicitly declared over the for loop. It only exists when the loop is executing, and not before or after it. For example, the following process statement shows how not to use the loop parameter:

```
erroneous : process is
    variable i, j : integer;
begin
    i := loop_param;                     -- error!
    for loop_param in 1 to 10 loop
        loop_param := 5;                 -- error!
    end loop;
    j := loop_param;                     -- error!
end process erroneous;
```

The assignments to i and j are illegal since the loop parameter is defined neither before nor after the loop. The assignment within the loop body is illegal because loop_param is a constant and thus may not be modified.

A consequence of the way the loop parameter is defined is that it *hides* any object of the same name defined outside the loop. For example, in this process:

```
hiding_example : process is
    variable a, b : integer;
begin
    a := 10;
    for a in 0 to 7 loop
        b := a;
    end loop;
    -- a = 10, and b = 7
    . . .
end process hiding_example;
```

the variable a is initially assigned the value 10, and then the for loop is executed, creating a loop parameter also called a. Within the loop, the assignment to b uses the loop parameter, so the final value of b after the last iteration is 7. After the loop, the loop parameter no longer exists, so if we use the name a, we are referring to the variable object, whose value is still 10.

As we mentioned above, the for loop iterates with the loop parameter assuming successive values from the discrete range starting from the leftmost value. An important point to note is that if we specify a null range, the for loop body does not execute at all. A null range can arise if we specify an ascending range with the left bound greater than the right bound, or a descending range with the left bound less than the right bound. For example, the for loop

```
for i in 10 to 1 loop
    . . .
end loop;
```

completes immediately, without executing the enclosed statements. If we really want the loop to iterate with i taking values 10, 9, 8 and so on, we should write

```
for i in 10 downto 1 loop
    . . .
end loop;
```

One final thing to note about for loops is that, like basic loop statements, they can enclose arbitrary sequential statements, including next and exit statements, and we can label a for loop by writing the label before the **for** keyword.

EXAMPLE

We now rewrite the cosine model in Figure 3-5 to calculate the result by summing the first 10 terms of the series. The entity declaration is unchanged. The revised architecture body, shown in Figure 3-6, consists of a process that uses a for loop instead of a while loop. As before, the variables sum and term are set to 1.0, representing the first term in the series. The variable n is replaced by the for

FIGURE 3-6

```
architecture fixed_length_series of cos is
begin
    summation : process (theta) is
        variable sum, term : real;
    begin
        sum := 1.0;
        term := 1.0;
        for n in 1 to 9 loop
            term := (–term) * theta**2 / real(((2*n–1) * 2*n));
            sum := sum + term;
        end loop;
        result <= sum;
    end process summation;
end architecture fixed_length_series;
```

The revised architecture body for the cosine module.

loop parameter. The loop iterates nine times, calculating the remaining nine terms
of the series.

Summary of Loop Statements

The preceding sections describe the various forms of loop statements in detail. It is
worth summarizing this information in one place, to show the few basic points to remember. First, the syntax rule for all loop statements is

loop_statement ⇐
    ⟦ *loop*_label : ⟧
    ⟦ **while** condition ‖ **for** identifier **in** discrete_range ⟧ **loop**
        { sequential_statement }
    end loop ⟦ *loop*_label ⟧ ;

Second, in the absence of exit and next statements, the while loop iterates as long
as the condition is true, and the for loop iterates with the loop parameter assuming
successive values from the discrete range. If the condition in a while loop is initially
false, or if the discrete range in a for loop is a null range, then no iterations occur.
    Third, the loop parameter in a for loop cannot be explicitly declared, and it is a
constant within the loop body. It also shadows any other object of the same name
declared outside the loop.
    Finally, an exit statement can be used to terminate any loop, and a next statement
can be used to complete the current iteration and commence the next iteration. These
statements can refer to loop labels to terminate or complete iteration for an outer level
of a nested set of loops.

3.5 Assertion and Report Statements

One of the reasons for writing models of computer systems is to verify that a design functions correctly. We can partially test a model by applying sample inputs and checking that the outputs meet our expectations. If they do not, we are then faced with the task of determining what went wrong inside the design. This task can be made easier using *assertion statements* that check that expected conditions are met within the model. An assertion statement is a sequential statement, so it can be included anywhere in a process body. The full syntax rule for an assertion statement is

> assertion_statement ⇐
>   ⟦ label : ⟧ **assert** *boolean*_expression
>   ⟦ **report** expression ⟧ ⟦ **severity** expression ⟧ ;

The optional label allows us to identify the assertion statement. The simplest form of assertion statement just includes the keyword **assert** followed by a condition expression that we expect to be true when the assertion statement is executed. If the condition is not met, we say that an *assertion violation* has occurred. If an assertion violation arises during simulation of a model, the simulator reports the fact. During synthesis, the condition in an assertion statement may be interpreted as a condition that the synthesizer may assume to be true. During formal verification, the condition may be interpreted as a condition to be proven by the verifier. For example, if we write

> **assert** initial_value <= max_value;

and initial_value is larger than max_value when the statement is executed during simulation, the simulator will let us know. During synthesis, the synthesizer may assume that initial_value <= max_value and optimize the circuit based on that information. During formal verification, the verifier may attempt to prove that initial_value <= max_value for all possible input stimuli and execution paths leading to the assertion statement.

If we have a number of assertion statements throughout a model, it is useful to know which assertion is violated. We can get the simulator to provide extra information by including a **report** clause in an assertion statement, for example:

> **assert** initial_value <= max_value
>   **report** "initial value too large";

The string that we provide is used to form part of the assertion violation message. We can write any expression in the report clause provided it yields a string value, for example:

> **assert** current_character >= '0' **and** current_character <= '9'
>   **report** "Input number " & input_string & " contains a non–digit";

Here the message is derived by concatenating three string values together.

In Section 2.2 on page 38, we mentioned a predefined enumeration type severity_level, defined as

> **type** severity_level **is** (note, warning, error, failure);

We can include a value of this type in a **severity** clause of an assertion statement. This value indicates the degree to which the violation of the assertion affects operation of the model. The value note can be used to pass informative messages out from a simulation, for example:

```
assert free_memory >= low_water_limit
    report "low on memory, about to start garbage collect"
    severity note;
```

The severity level warning can be used if an unusual situation arises in which the model can continue to execute, but may produce unusual results, for example:

```
assert packet_length /= 0
    report "empty network packet received"
    severity warning;
```

We can use the severity level error to indicate that something has definitely gone wrong and that corrective action should be taken, for example:

```
assert clock_pulse_width >= min_clock_width
    severity error;
```

Finally, the value failure can be used if we detect an inconsistency that should never arise, for example:

```
assert (last_position − first_position + 1) = number_of_entries
    report "inconsistency in buffer model"
    severity failure;
```

We have seen that we can write an assertion statement with either or both of a **report** clause and a **severity** clause. If both are present, the syntax rule shows us that the report clause must come first. If we omit the **report** clause, the default string in the error message is "Assertion violation." If we omit the **severity** clause, the default value is error. The severity value is usually used by a simulator to determine whether or not to continue execution after an assertion violation. Most simulators allow the user to specify a severity threshold, beyond which execution is stopped.

Usually, failure of an assertion means either that the entity is being used incorrectly as part of a larger design or that the model for the entity has been incorrectly written. We illustrate both cases.

EXAMPLE

A set/reset (SR) flipflop has two inputs, S and R, and an output Q. When S is '1', the output is set to '1', and when R is '1', the output is reset to '0'. However, S and R may not both be '1' at the same time. If they are, the output value is not specified. Figure 3-7 is a behavioral model for an SR flipflop that includes a check for this illegal condition.

The architecture body contains a process sensitive to the S and R inputs. Within the process body we write an assertion statement that requires that S and R not both be '1'. If both are '1', the assertion is violated, so the simulator writes an "Assertion violation" message with severity error. If execution continues after the violated assertion, the value '1' will first be assigned to Q, followed by the value

'0'. The resulting value is '0'. This is allowed, since the state of **Q** was not specified for this illegal condition, so we are at liberty to choose any value. If the assertion is not violated, then at most one of the following if statements is executed, correctly modeling the behavior of the SR flipflop.

FIGURE 3-7

```
entity SR_flipflop is
    port ( S, R : in bit;  Q : out bit );
end entity SR_flipflop;
_____

architecture checking of SR_flipflop is
begin
    set_reset : process (S, R) is
    begin
        assert S = '1' nand R = '1';
        if S = '1' then
            Q <= '1';
        end if;
        if R = '1' then
            Q <= '0';
        end if;
    end process set_reset;
end architecture checking;
```

An entity and architecture body for a set/reset flipflop, including a check for correct usage.

EXAMPLE

To illustrate the use of an assertion statement as a "sanity check," let us look at a model, shown in Figure 3-8, for an entity that has three integer inputs, **a, b** and **c**, and produces an integer output **z** that is the largest of its inputs.

The architecture body is written using a process containing nested if statements. For this example we have introduced an "accidental" error into the model. If we simulate this model and put the values **a** = 7, **b** = 3 and **c** = 9 on the ports of this entity, we expect that the value of result, and hence the output port, is 9. The assertion states that the value of result must be greater than or equal to all of the inputs. However, our coding error causes the value 7 to be assigned to result, and so the assertion is violated. This violation causes us to examine our model more closely, and correct the error.

FIGURE 3-8

```
entity max3 is
    port ( a, b, c : in integer;  z : out integer );
end entity max3;
```

(continued on page 76)

(continued from page 75)

```
architecture check_error of max3 is
begin
    maximizer : process (a, b, c)
        variable result : integer;
    begin
        if a > b then
            if a > c then
                result := a;
            else
                result := a;  -- Oops!  Should be: result := c;
            end if;
        elsif  b > c then
            result := b;
        else
            result := c;
        end if;
        assert result >= a and result >= b and result >= c
            report "inconsistent result for maximum"
            severity failure;
        z <= result;
    end process maximizer;
end architecture check_error;
```

An entity and architecture body for a maximum selector module, including a check for a correctly generated result.

Another important use for assertion statements is in checking timing constraints that apply to a model. For example, most clocked devices require that the clock pulse be longer than some minimum duration. We can use the predefined primary "now" in an expression to calculate durations. We return to "now" in a later chapter. Suffice it to say that it yields the current simulation time when it is evaluated.

EXAMPLE

An edge-triggered register has a data input and a data output of type **real** and a clock input of type **bit**. When the clock changes from '0' to '1', the data input is sampled, stored and transmitted through to the output. Let us suppose that the clock input must remain at '1' for at least 5 ns. Figure 3-9 is a model for this register, including a check for legal clock pulse width.

The architecture body contains a process that is sensitive to changes on the clock input. When the clock changes from '0' to '1', the input is stored, and the current simulation time is recorded in the variable **pulse_start**. When the clock changes from '1' to '0', the difference between **pulse_start** and the current simulation time is checked by the assertion statement.

FIGURE 3-9

```
entity edge_triggered_register is
    port ( clock : in bit;
                d_in : in real;  d_out : out real );
end entity edge_triggered_register;

- - - - - - - - - - - - - - - - - - - - - - - - - - - - - - - - - - - - - - - - - -

architecture check_timing of edge_triggered_register is
begin

    store_and_check : process (clock) is
        variable stored_value : real;
        variable pulse_start : time;
    begin
        case clock is
            when '1' =>
                pulse_start := now;
                stored_value := d_in;
                d_out <= stored_value;
            when '0' =>
                assert now = 0 ns or (now – pulse_start) >= 5 ns
                    report "clock pulse too short";
        end case;
    end process store_and_check;

end architecture check_timing;
```

An entity and architecture body for an edge-triggered register, including a timing check for correct pulse width on the clock input.

VHDL-87

Assertion statements may not be labeled in VHDL-87.

VHDL also provides us with a *report statement*, which is similar to an assertion statement. The syntax rule for the report statement shows this similarity:

report_statement ⇐
    〚 label : 〛 **report** expression 〚 **severity** expression 〛 ;

The differences are that there is no condition, and if the severity level is not specified, the default is **note**. Indeed, the report statement can be thought of as an assertion statement in which the condition is the value **false** and the severity is **note**, hence it always produces the message. One way in which the report statement is useful is as a means of including "trace writes" in a model as an aid to debugging.

EXAMPLE

Suppose we are writing a complex model and we are not sure that we have got the logic quite right. We can use report statements to get the processes in the model to write out messages, so that we can see when they are activated and what they are doing. An example process is

```
transmit_element : process (transmit_data) is
    . . .            − − variable declarations
begin
    report "transmit_element: data = "
               & data_type'image(transmit_data);

    . . .
end process transmit_element;
```

VHDL-87

Report statements are not provided in VHDL-87. We achieve the same effect by writing an assertion statement with the condition "**false**" and a severity level of note. For example, the VHDL-93 report statement

```
report "Initialization complete";
```

can be written in VHDL-87 as

```
assert false
    report "Initialization complete" severity note;
```

Exercises

1. [● 3.1] Write an if statement that sets a variable **odd** to '1' if an integer **n** is odd, or to '0' if it is even.

2. [● 3.1] Write an if statement that, given the year of today's date in the variable **year**, sets the variable **days_in_February** to the number of days in February. A year is a leap year if it is divisible by four, except for years that are divisible by 100.

3. [● 3.2] Write a case statement that strips the strength information from a standard-logic variable **x**. If **x** is '0' or 'L', set it to '0'. If **x** is '1' or 'H', set it to '1'. If **x** is 'X', 'W', 'Z', 'U' or '–', set it to 'X'. (This is the conversion performed by the standard-logic function **to_X01**.)

4. [● 3.2] Write a case statement that sets an integer variable **character_class** to 1 if the character variable **ch** contains a letter, to 2 if it contains a digit, to 3 if it contains some other printable character or to 4 if it contains a non-printable character. Note that the VHDL character set contains accented letters, as shown in Figure 2-2 on page 39.

5. [● 3.4] Write a loop statement that samples a bit input **d** when a clock input **clk** changes to '1'. So long as **d** is '0', the loop continues executing. When **d** is '1', the loop exits.

6. [❶ 3.4] Write a while loop that calculates the exponential function of x to an accuracy of one part in 10^4 by summing terms of the following series:

$$e^x = 1 + \frac{x}{1} + \frac{x^2}{2!} + \frac{x^3}{3!} + \frac{x^4}{4!} + \cdots$$

7. [❶ 3.4] Write a for loop that calculates the exponential function of x by summing the first eight terms of the series.

8. [❶ 3.5] Write an assertion statement that expresses the requirement that a flipflop's two outputs, q and q_n, of type std_ulogic, are complementary.

9. [❶ 3.5] We can use report statements in VHDL to achieve the same effect as using "trace writes" in software programming languages, to report a message when part of the model is executed. Insert a report statement in the model of Figure 3-4 to cause a trace message when the counter is reset.

10. [❷ 3.1] Develop a behavioral model for a limiter with three integer inputs, data_in, lower and upper; an integer output, data_out; and a bit output, out_of_limits. The data_out output follows data_in so long as it is between lower and upper. If data_in is less than lower, data_out is limited to lower. If data_in is greater than upper, data_out is limited to upper. The out_of_limit output indicates when data_out is limited.

11. [❷ 3.2] Develop a model for a floating-point arithmetic unit with data inputs x and y, data output z and function code input of an enumerated type with values add, sub, mult, div and recip. Function code add produces addition, sub produces subtraction of y from x, mult produces multiplication, div produces division of x by y and recip produces reciprocal of y.

12. [❷ 3.4] Write a model for a counter with an output port of type natural, initially set to 15. When the clk input changes to '1', the counter decrements by one. After counting down to zero, the counter wraps back to 15 on the next clock edge.

13. [❷ 3.4] Modify the counter of Exercise 12 to include an asynchronous load input and a data input. When the load input is '1', the counter is preset to the data input value. When the load input changes back to '0', the counter continues counting down from the preset value.

14. [❷ 3.4] Develop a model of an averaging module that calculates the average of batches of 16 real numbers. The module has clock and data inputs and a data output. The module accepts the next input number when the clock changes to '1'. After 16 numbers have been accepted, the module places their average on the output port, then repeats the process for the next batch.

15. [❷ 3.5] Write a model that causes assertion violations with different severity levels. Experiment with your simulator to determine its behavior when an assertion violation occurs. See if you can specify a severity threshold above which it stops execution.

Composite Data Types
and Operations

Now that we have seen the basic data types and sequential operations from which the behavioral part of a VHDL model is formed, it is time to look at composite data types. We first mentioned them in the classification of data types in Chapter 2. Composite data objects consist of related collections of data elements in the form of either an *array* or a *record*. We can treat an object of a composite type as a single object or manipulate its constituent elements individually. In this chapter, we see how to define composite types and how to manipulate them using operators and sequential statements.

4.1   Arrays

An *array* consists of a collection of values, all of which are of the same type as each other. The position of each element in an array is given by a scalar value called its *index*. To create an array object in a model, we first define an array type in a type declaration. The syntax rule for an array type definition is

> array_type_definition ⇐
>   **array** (discrete_range { , ₒₒₒ }) **of** *element*_subtype_indication

This defines an array type by specifying one or more index ranges (the list of discrete ranges) and the element type or subtype. Recall from previous chapters that a discrete range is a subset of values from a discrete type (an integer or enumeration type), and that it can be specified as shown by the simplified syntax rule

> discrete_range ⇐
>   *discrete*_subtype_indication
>   ‖ simple_expression (**to** ‖ **downto**) simple_expression

Recall also that a subtype indication can be just the name of a previously declared type (a type mark), and can include a range constraint to limit the set of values from that type, as shown by the simplified rule

> subtype_indication ⇐
>   type_mark ⟦ **range** simple_expression (**to** ‖ **downto**) simple_expression ⟧

We illustrate these rules for defining arrays with a series of examples. We start with single-dimensional arrays, in which there is just one index range. Here is a simple example to start off with, showing the declaration of an array type to represent words of data:

type word **is array** (0 **to** 31) **of** bit;

Each element is a bit, and the elements are indexed from 0 up to 31. An alternative declaration of a word type, more appropriate for "little-endian" systems, is

type word **is array** (31 **downto** 0) **of** bit;

The difference here is that index values start at 31 for the leftmost element in values of this type and continue down to 0 for the rightmost. The index values of an array do not have to be numeric. For example, given this declaration of an enumeration type:

type controller_state **is** (initial, idle, active, error);

we could then declare an array as follows:

type state_counts **is array** (idle **to** error) **of** natural;

This kind of array type declaration relies on the type of the index range being clear from the context. If there were more than one enumeration type with values idle and error, it would not be clear which one to use for the index type. To make it clear, we can use the alternative form for specifying the index range, in which we name the index type and include a range constraint. The previous example could be rewritten as

```
type state_counts is
    array (controller_state range idle to error) of natural;
```

If we need an array element for every value in an index type, we need only name the index type in the array declaration without specifying the range. For example:

```
subtype coeff_ram_address is integer range 0 to 63;
type coeff_array is array (coeff_ram_address) of real;
```

Once we have declared an array type, we can define objects of that type, including constants, variables and signals. For example, using the types declared above, we can declare variables as follows:

```
variable buffer_register, data_register : word;
variable counters : state_counts;
variable coeff : coeff_array;
```

Each of these objects consists of the collection of elements described by the corresponding type declaration. An individual element can be used in an expression or as the target of an assignment by referring to the array object and supplying an index value, for example:

```
coeff(0) := 0.0;
```

If active is a variable of type controller_state, we can write

```
counters(active) := counters(active) + 1;
```

An array object can also be used as a single composite object. For example, the assignment

```
data_register := buffer_register;
```

copies all of the elements of the array buffer_register into the corresponding elements of the array data_register.

EXAMPLE

Figure 4-1 is a model for a memory that stores 64 real-number coefficients, initialized to 0.0. We assume the type coeff_ram_address is previously declared as above. The architecture body contains a process with an array variable representing the coefficient storage. When the process starts, it initializes the array using a for loop. It then repetitively waits for any of the input ports to change. When rd is '1', the array is indexed using the address value to read a coefficient. When wr is '1', the address value is used to select which coefficient to change.

FIGURE 4-1

```
entity coeff_ram is
    port ( rd, wr : in bit;  addr : in coeff_ram_address;
            d_in : in real;  d_out : out real );
end entity coeff_ram;
```

(continued on page 84)

(continued from page 83)

```
architecture abstract of coeff_ram is
begin

    memory : process is
        type coeff_array is array (coeff_ram_address) of real;
        variable coeff : coeff_array;
    begin
        for index in coeff_ram_address loop
            coeff(index) := 0.0;
        end loop;
        loop
            wait on rd, wr, addr, d_in;
            if rd = '1' then
                d_out <= coeff(addr);
            end if;
            if wr = '1' then
                coeff(addr) := d_in;
            end if;
        end loop;
    end process memory;

end architecture abstract;
```

An entity and architecture body for a memory module that stores real-number coefficients. The memory storage is implemented using an array.

Multidimensional Arrays

VHDL also allows us to create multidimensional arrays, for example, to represent matrices or tables indexed by more than one value. A multidimensional array type is declared by specifying a list of index ranges, as shown by the syntax rule on page 82. For example, we might include the following type declarations in a model for a finite-state machine:

```
type symbol is ('a', 't', 'd', 'h', digit, cr, error);
type state is range 0 to 6;

type transition_matrix is array (state, symbol) of state;
```

Each index range can be specified as shown above for single-dimensional arrays. The index ranges for each dimension need not all be from the same type, nor have the same direction. An object of a multidimensional array type is indexed by writing a list of index values to select an element. For example, if we have a variable declared as

```
variable transition_table : transition_matrix;
```

we can index it as follows:

```
transition_table(5, 'd');
```

EXAMPLE
_____

In three-dimensional graphics, a point in space may be represented using a three-element vector [*x, y, z*] of coordinates. Transformations, such as scaling, rotation and reflection, may be done by multiplying a vector by a 3 × 3 transformation matrix to get a new vector representing the transformed point. We can write VHDL type declarations for points and transformation matrices:

```
type point is array (1 to 3) of real;
type matrix is array (1 to 3, 1 to 3) of real;
```

We can use these types to declare point variables p and q and a matrix variable transform:

```
variable p, q : point;
variable transform : matrix;
```

The transformation can be applied to the point p to produce a result in q with the following statements:

```
for i in 1 to 3 loop
    q(i) := 0.0;
    for j in 1 to 3 loop
        q(i) := q(i) + transform(i, j) * p(j);
    end loop;
end loop;
```
_____

Array Aggregates

We have seen how we can write literal values of scalar types. Often we also need to write literal array values, for example, to initialize a variable or constant of an array type. We can do this using a VHDL construct called an array *aggregate*, according to the syntax rule

$\quad$ aggregate $\Leftarrow$ (([choices =>] expression) { , ... })

Let us look first at the form of aggregate without the choices part. It simply consists of a list of the elements enclosed in parentheses, for example:

```
type point is array (1 to 3) of real;
constant origin : point := (0.0, 0.0, 0.0);
variable view_point : point := (10.0, 20.0, 0.0);
```

This form of array aggregate uses *positional association* to determine which value in the list corresponds to which element of the array. The first value is the element with the leftmost index, the second is the next index to the right, and so on, up to the last value, which is the element with the rightmost index. There must be a one-to-one correspondence between values in the aggregate and elements in the array.

An alternative form of aggregate uses *named association*, in which the index value for each element is written explicitly using the choices part shown in the syntax rule. The choices may be specified in exactly the same way as those in alternatives of a case statement, discussed in Chapter 3. As a reminder, here is the syntax rule for choices:

choices ⇐ (simple_expression ‖ discrete_range ‖ **others**) { | ○○○ }

For example, the variable declaration and initialization could be rewritten as

variable view_point : point := (1 => 10.0, 2 => 20.0, 3 => 0.0);

The main advantage of named association is that it gives us more flexibility in writing aggregates for larger arrays. To illustrate this, let us return to the coefficient memory described above. The type declaration was

type coeff_array **is array** (coeff_ram_address) **of** real;

Suppose we want to declare the coefficient variable, initialize the first few locations to some non-zero value and initialize the remainder to zero. Following are a number of ways of writing aggregates that all have the same effect:

variable coeff : coeff_array := (0 => 1.6, 1 => 2.3, 2 => 1.6, 3 **to** 63 => 0.0);

Here we are using a range specification to initialize the bulk of the array value to zero.

variable coeff : coeff_array := (0 => 1.6, 1 => 2.3, 2 => 1.6, **others** => 0.0);

The keyword **others** stands for any index value that has not been previously mentioned in the aggregate. If the keyword **others** is used, it must be the last choice in the aggregate.

variable coeff : coeff_array := (0 | 2 => 1.6, 1 => 2.3, **others** => 0.0);

The "|" symbol can be used to separate a list of index values, for which all elements have the same value.

Note that we may not mix positional and named association in an array aggregate, except for the use of an **others** choice in the final postion. Thus, the following aggregate is illegal:

variable coeff : coeff_array := (1.6, 2.3, 2 => 1.6, **others** => 0.0);   *-- illegal*

We can also use aggregates to write multidimensional array values. In this case, we treat the array as though it were an array of arrays, writing an array aggregate for each of the leftmost index values first.

EXAMPLE

We can use a two-dimensional array to represent the transition matrix of a finite-state machine (FSM) that interprets simple modem commands. A command must consist of the string "atd" followed by a string of digits and a **cr** character, or the string "ath" followed by **cr**. The state transition diagram is shown in Figure 4-2. The symbol "other" represents a character other than 'a', 't', 'd', 'h', a digit or **cr**. An outline of a process that implements the FSM is shown in Figure 4-3. The type declarations for **symbol** and **state** represent the command symbols and the states for the FSM. The transition matrix, **next_state**, is a two-dimensional array constant indexed by the state and symbol type. An element at position (i, j) in this matrix indicates the next state the FSM should move to when it is in state i and the next input symbol is j. The matrix is initialized according to the transition diagram. The process uses the **current_state** variable and successive input symbols

FIGURE 4-2

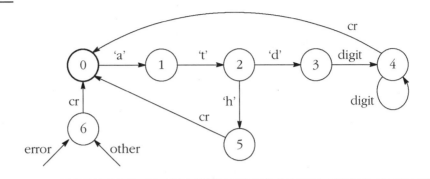

The state transition diagram for a modem command finite-state machine. State 0 is the initial state. The machine returns to this state after recognizing a correct command. State 6 is the error state, to which the machine goes if it detects an illegal or unexpected character.

FIGURE 4-3

```
modem_controller : process is
    type symbol is ('a', 't', 'd', 'h', digit, cr, other);
    type symbol_string is array (1 to 20) of symbol;
    type state is range 0 to 6;
    type transition_matrix is array (state, symbol) of state;

    constant next_state : transition_matrix :=
        ( 0 => ('a' => 1, others => 6),
          1 => ('t' => 2, others => 6),
          2 => ('d' => 3, 'h' => 5, others => 6),
          3 => (digit => 4, others => 6),
          4 => (digit => 4, cr => 0, others => 6),
          5 => (cr => 0, others => 6),
          6 => (cr => 0, others => 6) );

    variable command : symbol_string;
    variable current_state : state := 0;
begin
    . . .
    for index in 1 to 20 loop
        current_state := next_state( current_state, command(index) );
        case current_state is
            . . .
        end case;
    end loop;
    . . .
end process modem_controller;
```

An outline of a process that implements the finite-state machine to accept a modem command.

as indices into the transition matrix to determine the next state. For each transition, it performs some action based on the new state. The actions are implemented within the case statement.

Another place in which we may use an aggregate is the target of a variable assignment or a signal assignment. The full syntax rule for a variable assignment statement is

variable_assignment_statement ⇐
    〚 label : 〛 (name ‖ aggregate) := expression ;

If the target is an aggregate, it must contain a variable name at each element position. Furthermore, expression on the right-hand side of the assignment must produce a composite value of the same type as the target aggregate. Each element of the right-hand side is assigned to the corresponding variable in the target aggregate. The full syntax rule for a signal assignment also allows the target to be in the form of an aggregate, with each element being a signal name. We can use assignments of this form to split a composite value among a number of scalar signals. For example, if we have a variable **flag_reg**, which is a four-element bit vector, we can perform the following signal assignment to four signals of type **bit**:

    (z_flag, n_flag, v_flag, c_flag) <= flag_reg;

Since the right-hand side is a bit vector, the target is taken as a bit-vector aggregate. The leftmost element of **flag_reg** is assigned to **z_flag**, the second element of **flag_reg** is assigned to **n_flag**, and so on. This form of multiple assignment is much more compact to write than four separate assignment statements.

Array Attributes

In Chapter 2 we saw that attributes could be used to refer to information about scalar types. There are also attributes applicable to array types; they refer to information about the index ranges. Array attributes can also be applied to array objects, such as constants, variables and signals, to refer to information about the types of the objects. Given some array type or object **A**, and an integer **N** between 1 and the number of dimensions of **A**, VHDL defines the following attributes:

A'left(N)	Left bound of index range of dimension N of A
A'right(N)	Right bound of index range of dimension N of A
A'low(N)	Lower bound of index range of dimension N of A
A'high(N)	Upper bound of index range of dimension N of A
A'range(N)	Index range of dimension N of A
A'reverse_range(N)	Reverse of index range of dimension N of A
A'length(N)	Length of index range of dimension N of A
A'ascending(N)	true if index range of dimension N of A is an ascending range, false otherwise

For example, given the array declaration

    type A **is array** (1 **to** 4, 31 **downto** 0) **of** boolean;

some attribute values are

A'left(1) = 1	A'low(1) = 1
A'right(2) = 0	A'high(2) = 31
A'range(1) is 1 **to** 4	A'reverse_range(2) is 0 **to** 31
A'length(1) = 4	A'length(2) = 32
A'ascending(1) = true	A'ascending(2) = false

For all of these attributes, to refer to the first dimension (or if there is only one dimension), we can omit the dimension number in parentheses, for example:

    A'low = 1   A'length = 4

In the next section, we see how these array attributes may be used to deal with array ports. We will also see, in Chapter 6, how they may be used with subprogram parameters that are arrays. Another major use is in writing for loops to iterate over elements of an array. For example, given an array variable **free_map** that is an array of bits, we can write a for loop to count the number of '1' bits without knowing the actual size of the array:

```
count := 0;
for index in free_map'range loop
    if free_map(index) = '1' then
        count := count + 1;
    end if;
end loop;
```

The 'range and 'reverse_range attributes can be used in any place in a VHDL model where a range specification is required, as an alternative to specifying the left and right bounds and the range direction. Thus, we may use the attributes in type and subtype definitions, in subtype constraints, in for loop parameter specifications, in case statement choices and so on. The advantage of taking this approach is that we can specify the size of the array in one place in the model and in all other places use array attributes. If we need to change the array size later for some reason, we need only change the model in one place.

VHDL-87

The array attribute 'ascending is not provided in VHDL-87.

4.2 Unconstrained Array Types

The array types we have seen so far in this section are called *constrained* arrays, since the type definition constrains index values to be within a specific range. VHDL also allows us to define *unconstrained* array types, in which we just indicate the type of the index values, without specifying bounds. An unconstrained array type definition is described by the alternate syntax rule

> array_type_definition ⇐
>   **array** (⟨ type_mark **range** <>) { , ₒₒₒ })
>   **of** *element*_subtype_indication

The symbol "<>", often called "box," can be thought of as a placeholder for the index range, to be filled in later when the type is used. An example of an unconstrained array type declaration is

> **type** sample **is array** (natural **range** <>) **of** integer;

An important point to understand about unconstrained array types is that when we declare an object of such a type, we need to provide a constraint that specifies the index bounds. We can do this in several ways. One way is to provide the constraint when an object is created, for example:

> **variable** short_sample_buf : sample(0 **to** 63);

This indicates that index values for the variable **short_sample** are natural numbers in the ascending range 0 to 63. Another way to specify the constraint is to declare a subtype of the unconstrained array type. Objects can then be created using this sub-type, for example:

> **subtype** long_sample **is** sample(0 **to** 255);
> **variable** new_sample_buf, old_sample_buf : long_sample;

These are both examples of a new form of subtype indication that we have not yet seen. The syntax rule is

> subtype_indication ⇐ type_mark ⟦ (discrete_range { , ₒₒₒ }) ⟧

The type mark is the name of the unconstrained array type, and the discrete range spec-ifications constrain the index type to a subset of values used to index array elements. Each discrete range must be of the same type as the corresponding index type.

When we declare a constant of an unconstrained array type, there is a third way in which we can provide a constraint. We can infer it from the expression used to ini-tialize the constant. If the initialization expression is an array aggregate written using named association, the index values in the aggregate imply the index range of the constant. For example, in the constant declaration

> **constant** lookup_table : sample := (1 => 23, 3 => −16, 2 => 100, 4 => 11);

the index range is 1 to 4.

If the expression is an aggregate using positional association, the index value of the first element is assumed to be the leftmost value in the array subtype. For example, in the constant declaration

> **constant** beep_sample : sample := (127, 63, 0, –63, –127, –63, 0, 63);

the index range is 0 to 7, since the index subtype is natural. The index direction is ascending, since natural is defined to be an ascending range.

Strings

VHDL provides a predefined unconstrained array type called string, declared as

> **type** string **is array** (positive **range** <>) **of** character;

In principle the index range for a constrained string may be either an ascending or descending range, with any positive integers for the index bounds. However, most applications simply use an ascending range starting from 1. For example:

> **constant** LCD_display_len : positive := 20;
> **subtype** LCD_display_string **is** string(1 **to** LCD_display_len);
> **variable** LCD_display : LCD_display_string := (**others** => ' ');

Bit Vectors

VHDL also provides a predefined unconstrained array type called bit_vector, declared as

> **type** bit_vector **is array** (natural **range** <>) **of** bit;

This type can be used to represent words of data at the architectural level of modeling. For example, subtypes for representing bytes of data in a little-endian processor might be declared as

> **subtype** byte **is** bit_vector(7 **downto** 0);

Alternatively, we can supply the constraint when an object is declared, as in the following:

> **variable** channel_busy_register : bit_vector(1 **to** 4);

Standard-Logic Arrays

The standard-logic package std_logic_1164 provides an unconstrained array type for vectors of standard-logic values. It is declared as

> **type** std_ulogic_vector **is array** (natural **range** <>) **of** std_ulogic;

This type can be used in a way similar to bit vectors, but provides more detail in representing the electrical levels used in a design. We can define subtypes of the standard-logic vector type, for example:

> **subtype** std_ulogic_word **is** std_ulogic_vector(0 **to** 31);

Or we can directly create an object of the standard-logic vector type:

signal csr_offset : std_ulogic_vector(2 **downto** 1);

String and Bit-String Literals

In Chapter 1, we saw that a string literal may be used to write a value representing a sequence of characters. We can use a string literal in place of an array aggregate for a value of type string. For example, we can initialize a string constant as follows:

constant ready_message : string := "Ready   ";

We can also use string literals for any other one-dimensional array type whose elements are of an enumeration type that includes characters. The IEEE standard-logic array type std_ulogic_vector is an example. Thus we could declare and initialize a variable as follows:

variable current_test : std_ulogic_vector(0 **to** 13) := "ZZZZZZZZZZ————";

In Chapter 1 we also saw bit-string literals as a way of writing a sequence of bit values. Bit strings can be used in place of array aggregates to write values of bit-vector types. For example, the variable channel_busy_register defined above may be initialized with an assignment:

channel_busy_register := b"0000";

We can also use bit-string literals for other one-dimensional array types whose elements are of an enumeration type that includes the characters '0' and '1'. Each character in the bit-string literal represents one, three or four successive elements of the array value, depending on whether the base specified in the literal is binary, octal or hexadecimal. Again, using std_ulogic_vector as an example type, we can write a constant declaration using a bit-string literal:

constant all_ones : std_ulogic_vector(15 **downto** 0) := X"FFFF";

VHDL-87

Bit-string literals may only be used as literals for array types in which the elements are of type bit. The predefined type bit_vector is such a type. However, the standard-logic type std_ulogic_vector is not. We may use string literals for array types such as std_ulogic_vector.

Unconstrained Array Ports

An important use of an unconstrained array type is to specify the type of an array port. This use allows us to write an entity interface in a general way, so that it can connect to array signals of any size or with any range of index values. When we instantiate the entity, the index bounds of the array signal connected to the port are used as the bounds of the port.

EXAMPLE

Suppose we wish to model a family of and gates, each with a different number of inputs. We declare the entity interface as shown in Figure 4-4. The input port is of the unconstrained type bit_vector. The architecture body includes a process that is sensitive to changes on the input port. When any element changes, the process performs a logical and operation across the input array. It uses the 'range attribute to determine the index range of the array, since the index range is not known until the entity is instantiated.

FIGURE 4-4

```
entity and_multiple is
    port ( i : in bit_vector;  y : out bit );
end entity and_multiple;
- - - - - - - - - - - - - - - - - - - - - - - - - - - - - - - - - - - - - - - -
architecture behavioral of and_multiple is
begin
    and_reducer : process ( i ) is
        variable result : bit;
    begin
        result := '1';
        for index in i'range loop
            result := result and i(index);
        end loop;
        y <= result;
    end process and_reducer;
end architecture behavioral;
```

An entity and architecture body for an and gate with an unconstrained array input port.

To illustrate the use of the multiple-input gate entity, suppose we have the following signals:

```
signal count_value : bit_vector(7 downto 0);
signal terminal_count : bit;
```

We instantiate the entity, connecting its input port to the bit-vector signal:

```
tc_gate : entity work.and_multiple(behavioral)
    port map ( i => count_value, y => terminal_count);
```

For this instance, the input port is constrained by the index range of the signal. The instance acts as an eight-input and gate.

4.3 Array Operations and Referencing

Although an array is a collection of values, much of the time we operate on arrays one element at a time, using the operators described in Chapter 2. However, if we are working with one-dimensional arrays of scalar values, we can use some of the operators to operate on whole arrays, combining elements in a pairwise fashion.

First, the logical operators (**and, or, nand, nor, xor** and **xnor**) can be applied to two one-dimensional arrays of bit or Boolean elements. The operands must be of the same length and type, and the result is computed by applying the operator to matching elements from each array to produce an array of the same length. Elements are matched starting from the leftmost position in each array. An element at a given position from the left in one array is matched with the element at the same position from the left in the other array. The operator **not** can also be applied to a single array of bit or Boolean elements, with the result being an array of the same length and type as the operand. The following declarations and statements illustrate this use of logical operators when applied to bit vectors:

```
subtype pixel_row is bit_vector (0 to 15);
variable current_row, mask : pixel_row;

current_row := current_row and not mask;
current_row := current_row xor X"FFFF";
```

Second, the shift operators introduced in Chapter 2 (**sll, srl, sla, sra, rol** and **ror**) can be used with a one-dimensional array of bit or Boolean values as the left operand and an integer value as the right operand. A shift-left logical operation shifts the elements in the array n places to the left (n being the right operand), filling in the vacated positions with '0' or **false** and discarding the leftmost n elements. If n is negative, the elements are instead shifted to the right. Some examples are

> B"10001010" **sll** 3 = B"01010000"   B"10001010" **sll** −2 = B"00100010"

The shift-right logical operation similarly shifts elements n positions to the right for positive n, or to the left for negative n, for example:

> B"10010111" **srl** 2 = B"00100101"   B"10010111" **srl** −6 = B"11000000"

The next two shift operations, shift-left arithmetic and shift-right arithmetic, operate similarly, but instead of filling vacated positions with '0' or **false**, they fill them with a copy of the element at the end being vacated, for example:

> B"01001011" **sra** 3 = B"00001001"   B"10010111" **sra** 3 = B"11110010"
> B"00001100" **sla** 2 = B"00110000"   B"00010001" **sla** 2 = B"01000111"

As with the logical shifts, if n is negative, the shifts work in the opposite direction, for example:

> B"00010001" **sra** −2 = B"01000111"
> B"00110000" **sla** −2 = B"00001100"

A rotate-left operation moves the elements of the array n places to the left, transferring the n elements from the left end of the array around to the vacated positions at the right end. A rotate-right operation does the same, but in the opposite direction.

As with the shift operations, a negative right argument reverses the direction of rotation. Some examples are

$$B"10010011" \textbf{ rol } 1 = B"00100111" \qquad B"10010011" \textbf{ ror } 1 = B"11001001"$$

Relational operators form the third group of operations that can be applied to one-dimensional arrays. The array elements can be of any discrete type. The two operands need not be of the same length, so long as they have the same element type. The way these operators work can be most easily seen when they are applied to strings of characters, in which case they are compared according to case-sensitive dictionary ordering.

To see how dictionary comparison can be generalized to one-dimensional arrays of other element types, let us consider the "<" operator applied to two arrays a and b. If both a and b have length 0, a < b is false. If a has length 0, and b has non-zero length, then a < b. Alternatively, if both a and b have non-zero length, then a < b if a(1) < b(1), or if a(1) = b(1) and the rest of a < the rest of b. In the remaining case, where a has non-zero length and b has length 0, a < b is false. Comparison using the other relational operators is performed analogously.

The one remaining operator that can be applied to one-dimensional arrays is the concatenation operator (&), which joins two array values end to end. For example, when applied to bit vectors, it produces a new bit vector with length equal to the sum of the lengths of the two operands. Thus, b"0000" & b"1111" produces b"0000_1111".

The concatenation operator can be applied to two operands, one of which is an array and the other of which is a single scalar element. It can also be applied to two scalar values to produce an array of length 2. Some examples are

```
"abc" & 'd' = "abcd"
'w' & "xyz" = "wxyz"
'a' & 'b' = "ab"
```

VHDL-87

The logical operator **xnor** and the shift operators **sll**, **srl**, **sla**, **sra**, **rol** and **ror** are not provided in VHDL-87.

Array Slices

Often we want to refer to a contiguous subset of elements of an array, but not the whole array. We can do this using *slice* notation, in which we specify the left and right index values of part of an array object. For example, given arrays a1 and a2 declared as follows:

```
type array1 is array (1 to 100) of integer;
type array2 is array (100 downto 1) of integer;

variable a1 : array1;
variable a2 : array2;
```

we can refer to the array slice a1(11 to 20), which is an array of 10 elements having the indices 11 to 20. Similarly, the slice a2(50 downto 41) is an array of 10 elements but with a descending index range. Note that the slices a1(10 **to** 1) and a2(1 **downto** 10) are *null* slices, since the index ranges specified are null. Furthermore, the ranges specified in the slice must have the same direction as the original array. Thus we may not legally write a1(10 **downto** 1) or a2(1 **to** 10).

EXAMPLE

Figure 4-5 is a behavioral model for a byte-swapper that has one input port and one output port, each of which is a bit vector of subtype **halfword**, declared as follows:

subtype halfword **is** bit_vector(0 **to** 15);

The process in the architecture body swaps the two bytes of input with each other. It shows how the slice notation can be used for signal array objects in signal assignment statements.

FIGURE 4-5

```
entity byte_swap is
    port ( input : in halfword;  output : out halfword );
end entity byte_swap;

architecture behavior of byte_swap is
begin
    swap : process (input)
    begin
        output(8 to 15) <= input(0 to 7);
        output(0 to 7) <= input(8 to 15);
    end process swap;
end architecture behavior;
```

An entity and architecture body for a byte-swapper module.

VHDL-87

In VHDL-87, the range specified in a slice may have the opposite direction to that of the index range of the array. In this case, the slice is a null slice.

Array Type Conversions

In Chapter 2 we introduced the idea of type conversion of a numeric value to another value of a closely related type. A value of an array type can also be converted to a value of another array type, provided both array types have the same element type, the same number of dimensions and index types that can be type converted. The type conversion simply produces a new array value of the specified type, with each index converted to the value in the corresponding position of the new type's index range.

To illustrate the idea of type-converting array values, suppose we have the following declarations in a model:

```
subtype name is string(1 to 20);
type display_string is array (integer range 0 to 19) of character;

variable item_name : name;
variable display : display_string;
```

We cannot directly assign the value of item_name to display, since the types are different. However, we can using a type conversion:

```
display := display_string(item_name);
```

This produces a new array, with the left element having index 0 and the right element having index 19, which is compatible with the assignment target.

A common case in which we do not need a type conversion is the assigment of an array value of one subtype to an array object of a different subtype of the same base type. This occurs where the index ranges of the target and the operand have different bounds or directions. VHDL automatically includes an implicit subtype conversion in the assignment. For example, given the subtypes and variables declared thus:

```
subtype big_endian_upper_halfword is bit_vector(0 to 15);
subtype little_endian_upper_halfword is bit_vector(31 downto 16);

variable big : big_endian_upper_halfword;
variable little : little_endian_upper_halfword;
```

we could make the following assignments without including explicit type conversions:

```
big := little;
little := big;
```

4.4 Records

In this section, we discuss the second class of composite types, *records*. A record is a composite value comprising elements that may be of different types from one another. Each element is identified by a name, which is unique within the record. This name is used to select the element from the record value. The syntax rule for a record type definition is

```
record_type_definition ⇐
    record
        ( identifier { , ₀₀₀ } : subtype_indication ; )
        { ₀₀₀ }
    end record ⟦ identifier ⟧
```

Each of the names in the identifier lists declares an element of the indicated type or subtype. Recall that the curly brackets in the syntax rule indicate that the enclosed part may be repeated indefinitely. Thus, we can include several elements of different types within the record. The identifier at the end of the record type definition, if included, must repeat the name of the record type.

VHDL-87

The record type name may not be included at the end of a record type definition in VHDL-87.

The following is an example record type declaration and variable declarations using the record type:

```
type time_stamp is record
        seconds : integer range 0 to 59;
        minutes : integer range 0 to 59;
        hours : integer range 0 to 23;
    end record time_stamp;
variable sample_time, current_time : time_stamp;
```

Whole record values can be assigned using assignment statements, for example:

```
sample_time := current_time;
```

We can also refer to an individual element in a record using a *selected name*, for example:

```
sample_hour := sample_time.hours;
```

In the expression on the right of the assignment symbol, the prefix before the dot names the record value, and the suffix after the dot selects the element from the record. A selected name can also be used on the left side of an assignment to identify a record element to be modified, for example:

```
current_time.seconds := clock mod 60;
```

EXAMPLE

In the early stages of designing a new instruction set for a CPU, we don't want to commit to an encoding of opcodes and operands within an instruction word. Instead we use a record type to represent the components of an instruction. We illustrate this in Figure 4-6, an outline of a system-level behavioral model of a CPU and memory that uses record types to represent instructions and data.

FIGURE 4-6

```
architecture system_level of computer is
    type opcodes is (add, sub, addu, subu, jmp, breq, brne, ld, st, . . . );
    type reg_number is range 0 to 31;
    constant r0 : reg_number := 0;  constant r1 : reg_number := 1;  . . .
    type instruction is record
            opcode : opcodes;
            source_reg1, source_reg2, dest_reg : reg_number;
            displacement : integer;
        end record instruction;
```

```vhdl
        type word is record
                instr : instruction;
                data : bit_vector(31 downto 0);
            end record word;

    signal address : natural;
    signal read_word, write_word : word;
    signal mem_read, mem_write : bit := '0';
    signal mem_ready : bit := '0';

begin

    cpu : process is
        variable instr_reg : instruction;
        variable PC : natural;
        . . .        – – other declarations for register file, etc.
    begin
        address <= PC;
        mem_read <= '1';
        wait until mem_ready = '1';
        instr_reg := read_word.instr;
        mem_read <= '0';
        PC := PC + 4;
        case instr_reg.opcode is        – – execute the instruction
            . . .
        end case;
    end process cpu;

    memory : process is
        type memory_array is array (0 to 2**14 – 1) of word;
        variable store : memory_array :=
            ( 0  => ( ( ld, r0, r0, r2, 40 ), X"00000000" ),
              1  => ( ( breq, r2, r0, r0, 5 ), X"00000000" ),
              . . .
              40  => ( ( nop, r0, r0, r0, 0 ), X"FFFFFFFE"),
              others => ( ( nop, r0, r0, r0, 0 ), X"00000000") );
    begin
        . . .
    end process memory;
end architecture system_level;
```

An outline of a behavioral architecture body for a computer system comprising a CPU and a memory, using record values to represent instructions and data values.

The record type **instruction** represents the information to be included in each instruction of a program and includes the opcode, source and destination register numbers and a displacement. The record type **word** represents a word stored in memory. Since a word might represent an instruction or data, elements are included in the record for both possibilities. Unlike many conventional programming languages, VHDL does not provide variant parts in record values. The record type **word** illustrates how composite data values can include elements that are themselves composite values, provided the included elements are of a constrained subtype. The signals in the model are used for the address, data and control connections between the CPU and the memory.

Within the CPU process the variable instr_reg represents the instruction register containing the current instruction to be executed. The process fetches a word from memory and copies the instruction element from the record into the instruction register. It then uses the opcode field of the value to determine how to execute the instruction.

The memory process contains a variable that is an array of word records representing the memory storage. The array is initialized with a program and data. Words representing instructions are initialized with a record aggregate containing an instruction record aggregate and a bit vector, which is ignored. Similarly, words representing data are initialized with an aggregate containing an instruction aggregate, which is ignored, and the bit vector of data.

Record Aggregates

We can use a record aggregate to write a literal value of a record type—for example, to initialize a record variable or constant. Using a record aggregate is analogous to using an array aggregate for writing a literal value of an array type (see page 85). A record aggregate is formed by writing a list of the elements enclosed in parentheses. An aggregate using positional association lists the elements in the same order as they appear in the record type declaration. For example, given the record type time_stamp shown above, we can initialize a constant as follows:

    constant midday : time_stamp := (0, 0, 12);

We can also use named association, in which we identify each element in the aggregate by its name. The order of elements identified using named association does not affect the aggregate value. The example above could be rewritten as

    constant midday : time_stamp := (hours => 12, minutes => 0, seconds => 0);

Unlike array aggregates, we can mix positional and named association in record aggregates, provided all of the named elements follow any positional elements. We can also use the symbols " | " and **others** when writing choices. Here are some more examples, using the types instruction and time_stamp declared above:

```
constant nop_instr : instruction :=
        ( opcode => addu,
          source_reg1 | source_reg2 | dest_reg => 0,
          displacement => 0 );
```

    variable latest_event : time_stamp := (**others** => 0); – – *initially midnight*

Note that unlike array aggregates, we can't use a range of values to identify elements in a record aggregate, since the elements are identified by names, not indexed by a discrete range.

Exercises

1. [● 4.1] Write an array type declaration for an array of 30 integers, and a variable declaration for a variable of the type. Write a for loop to calculate the average of the array elements.

2. [● 4.1] Write an array type declaration for an array of bit values, indexed by standard-logic values. Then write a declaration for a constant, std_ulogic_to_bit, of this type, that maps standard-logic values to the corresponding bit value. (Assume unknown values map to '0'.) Given a standard-logic vector v1 and a bit-vector variable v2, both indexed from 0 to 15, write a for loop that uses the constant std_ulogic_to_bit to map the standard-logic vector to the bit vector.

3. [● 4.1] The data on a diskette is arranged in 18 sectors per track, 80 tracks per side and two sides per diskette. A computer system maintains a map of free sectors. Write a three-dimensional array type declaration to represent such a map, with a '1' element representing a free sector and a '0' element representing an occupied sector. Write a set of nested for loops to scan a variable of this type to find the location of the first free sector.

4. [● 4.2] Write a declaration for a subtype of std_ulogic_vector, representing a byte. Declare a constant of this subtype, with each element having the value 'Z'.

5. [● 4.2] Write a for loop to count the number of '1' elements in a bit-vector variable v.

6. [● 4.3] An eight-bit vector v1 representing a two's-complement binary integer can be sign-extended into a 32-bit vector v2 by copying it to the leftmost eight positions of v2, then performing an arithmetic right shift to move the eight bits to the rightmost eight positions. Write variable assignment statements that use slicing and shift operations to express this procedure.

7. [● 4.4] Write a record type declaration for a test stimulus record containing a stimulus bit vector of three bits, a delay value and an expected response bit vector of eight bits.

8. [❷ 4.1] Develop a model for a register file that stores 16 words of 32 bits each. The register file has data input and output ports, each of which is a 32-bit word; read-address and write-address ports, each of which is an integer in the range 0 to 15; and a write-enable port of type bit. The data output port reflects the content of the location whose address is given by the read-address port. When the write-enable port is '1', the input data is written to the register file at the location whose address is given by the write-address port.

9. [❷ 4.1] Develop a model for a priority encoder with a 16-element bit-vector input port, an output port of type natural that encodes the index of the leftmost '1' value in the input and an output of type bit that indicates whether any input elements are '1'.

10. [❷ 4.2] Write a package that declares an unconstrained array type whose elements are integers. Use the type in an entity declaration for a module that finds the maximum of a set of numbers. The entity has an input port of the unconstrained array type and an integer output. Develop a behavioral architecture body for the entity. How should the module behave if the actual array associated with the input port is empty (i.e., of zero length)?

11. [❷ 4.2/4.3] Develop a model for a general and-or-invert gate, with two standard-logic vector input ports **a** and **b** and a standard-logic output port **y**. The output of the gate is $\overline{a_0 . b_0 + a_1 . b_1 + \cdots + a_{n-1} . b_{n-1}}$.

12. [❷ 4.4] Develop a model of a 3-to-8 decoder and a test bench to exercise the decoder. In the test bench, declare the record type that you wrote for Exercise 7 and a constant array of test record values. Initialize the array to a set of test vectors for the decoder, and use the vectors to perform the test.

Basic Modeling Constructs 5

The description of a module in a digital system can be divided into two facets: the external view and the internal view. The external view describes the interface to the module, including the number and types of inputs and outputs. The internal view describes how the module implements its function. In VHDL, we can separate the description of a module into an *entity declaration*, which describes the external interface, and one or more *architecture bodies*, which describe alternative internal implementations. These were introduced in Chapter 1 and are discussed in detail in this chapter. We also look at how a design is processed in preparation for simulation or synthesis.

5.1   Entity Declarations

Let us first examine the syntax rules for an entity declaration and then show some examples. We start with a simplified description of entity declarations and move on to a full description later in this chapter. The syntax rules for this simplified form of entity declaration are

> entity_declaration ⇐
>   **entity** identifier **is**
>   〚 **port** (*port*_interface_list) ; 〛
>   { entity_declarative_item }
>   **end** 〚 **entity** 〛 〚 identifier 〛 ;
> interface_list ⇐
>   (identifier { , ... } : 〚 mode 〛 subtype_indication
>   〚 := expression 〛) { ; ... }
>
> mode ⇐ **in** ‖ **out** ‖ **inout**

The identifier in an entity declaration names the module so that it can be referred to later. If the identifier is included at the end of the declaration, it must repeat the name of the entity. The port clause names each of the *ports*, which together form the interface to the entity. We can think of ports as being analogous to the pins of a circuit; they are the means by which information is fed into and out of the circuit. In VHDL, each port of an entity has a *type*, which specifies the kind of information that can be communicated, and a *mode*, which specifies whether information flows into or out from the entity through the port. These aspects of type and direction are in keeping with the strong typing philosophy of VHDL, which helps us avoid erroneous circuit descriptions. A simple example of an entity declaration is

```
entity adder is
    port ( a : in word;
           b : in word;
           sum : out word );
end entity adder;
```

This example describes an entity named **adder**, with two input ports and one output port, all of type **word**, which we assume is defined elsewhere. We can list the ports in any order; we do not have to put inputs before outputs. Also, we can include a list of ports of the same mode and type instead of writing them out individually. Thus the above declaration could equally well be written as follows:

```
entity adder is
    port ( a, b : in word;
           sum : out word );
end entity adder;
```

In this example we have seen input and output ports. We can also have bidirectional ports, with mode **inout**. These can be used to model devices that alternately sense and drive data through a pin. Such models must deal with the possibility of more than one connected device driving a given signal at the same time. VHDL provides a mechanism for this, *signal resolution*, which we will return to in Chapter 8.

The similarity between the description of a port in an entity declaration and the declaration of a variable may be apparent. This similarity is not coincidental, and we can extend the analogy by specifying a default value on a port description, for example:

```
entity and_or_inv is
    port ( a1, a2, b1, b2 : in bit := '1';
            y : out bit );
end entity and_or_inv;
```

The default value, in this case the '1' on the input ports, indicates the value each port should assume if it is left unconnected in an enclosing model. We can think of it as describing the value that the port "floats to." On the other hand, if the port is used, the default value is ignored. We say more about use of default values when we look at the execution of a model.

Another point to note about entity declarations is that the port clause is optional. So we can write an entity declaration such as

```
entity top_level is
end entity top_level;
```

which describes a completely self-contained module. As the name in this example implies, this kind of module usually represents the top level of a design hierarchy.

Finally, if we return to the first syntax rule on page 104, we see that we can include declarations of items within an entity declaration. These include declarations of constants, types, signals and other kinds of items that we will see later in this chapter. The items can be used in all architecture bodies corresponding to the entity. Thus, it makes sense to include declarations that are relevant to the entity and all possible implementations. Anything that is part of only one particular implementation should instead be declared within the corresponding architecture body.

EXAMPLE

Suppose we are designing an embedded controller using a microprocessor with a program stored in a read-only memory (ROM). The program to be stored in the ROM is fixed, but we still need to model the ROM at different levels of detail. We can include declarations that describe the program in the entity declaration for the ROM, as shown in Figure 5-1. These declarations are not directly accessible to a user of the ROM entity, but serve to document the contents of the ROM. Each architecture body corresponding to the entity can use the constant **program** to initialize whatever structure it uses internally to implement the ROM.

FIGURE 5-1

```
entity program_ROM is
    port ( address : in std_ulogic_vector(14 downto 0);
            data : out std_ulogic_vector(7 downto 0);
            enable : in std_ulogic );
```

(continued on page 106)

(continued from page 105)

```
        subtype instruction_byte is bit_vector(7 downto 0);
        type program_array is array (0 to 2**14 – 1) of instruction_byte;
        constant program : program_array
            := ( X"32", X"3F", X"03",    – – LDA  $3F03
                 X"71", X"23",           – – BLT    $23
              . . .
                 );
    end entity program_ROM;
```

An entity declaration for a ROM, including declarations that describe the program contained in it.

VHDL-87

The keyword **entity** may not be included at the end of an entity declaration in VHDL-87.

5.2 Architecture Bodies

The internal operation of a module is described by an architecture body. An architecture body generally applies some operations to values on input ports, generating values to be assigned to output ports. The operations can be described either by processes, which contain sequential statements operating on values, or by a collection of components representing sub-circuits. Where the operation requires generation of intermediate values, these can be described using *signals*, analogous to the internal wires of a module. The syntax rule for architecture bodies shows the general outline:

```
architecture_body ⇐
    architecture identifier of entity_name is
        { block_declarative_item }
    begin
        { concurrent_statement }
    end [ architecture ] [ identifier ] ;
```

The identifier names this particular architecture body, and the entity name specifies which module has its operation described by this architecture body. If the identifier is included at the end of the architecture body, it must repeat the name of the architecture body. There may be several different architecture bodies corresponding to a single entity, each describing an alternative way of implementing the module's operation. The block declarative items in an architecture body are declarations needed to implement the operations. The items may include type and constant declarations, signal declarations and other kinds of declarations that we will look at in later chapters.

VHDL-87

The keyword **architecture** may not be included at the end of an architecture body in VHDL-87.

Concurrent Statements

The *concurrent statements* in an architecture body describe the module's operation. One form of concurrent statement, which we have already seen, is a process statement. Putting this together with the rule for writing architecture bodies, we can look at a simple example of an architecture body corresponding to the **adder** entity on page 104:

```
architecture abstract of adder is
begin
    add_a_b : process (a, b) is
    begin
        sum <= a + b;
    end process add_a_b;
end architecture abstract;
```

The architecture body is named **abstract**, and it contains a process **add_a_b**, which describes the operation of the entity. The process assumes that the operator "+" is defined for the type word, the type of **a** and **b**. We will see in Chapter 6 how such a definition may be written. We could also envisage additional architecture bodies describing the adder in different ways, provided they all conform to the external interface laid down by the entity declaration.

We have looked at processes first because they are the most fundamental form of concurrent statement. All other forms can ultimately be reduced to one or more processes. Concurrent statements are so called because conceptually they can be activated and perform their actions together, that is, concurrently. Contrast this with the sequential statements inside a process, which are executed one after another. Concurrency is useful for modeling the way real circuits behave. If we have two gates whose inputs change, each evaluates its new output independently of the other. There is no inherent sequencing governing the order in which they are evaluated. We look at process statements in more detail in Section 5.3. Then, in Section 5.4, we look at another form of concurrent statement, the component instantiation statement, used to describe how a module is composed of interconnected sub-modules.

Signal Declarations

When we need to provide internal signals in an architecture body, we must define them using *signal declarations*. The syntax for a signal declaration is very similar to that for a variable declaration:

signal_declaration ⇐
    signal identifier { , ... } : subtype_indication ⟦ := expression ⟧ ;

This declaration simply names each signal, specifies its type and optionally includes an initial value for all signals declared by the declaration.

EXAMPLE

Figure 5-2 is an example of an architecture body for the entity **and_or_inv**, defined on page 105. The architecture body includes declarations of some signals that are internal to the architecture body. They can be used by processes within the architecture body but are not accessible outside, since a user of the module need not be concerned with the internal details of its implementation. Values are assigned to signals using signal assignment statements within processes. Signals can be sensed by processes to read their values.

FIGURE 5-2

```
architecture primitive of and_or_inv is
    signal and_a, and_b : bit;
    signal or_a_b : bit;
begin
    and_gate_a : process (a1, a2) is
    begin
        and_a <= a1 and a2;
    end process and_gate_a;
    and_gate_b : process (b1, b2) is
    begin
        and_b <= b1 and b2;
    end process and_gate_b;
    or_gate : process (and_a, and_b) is
    begin
        or_a_b <= and_a or and_b;
    end process or_gate;
    inv : process (or_a_b) is
    begin
        y <= not or_a_b;
    end process inv;
end architecture primitive;
```

An architecture body corresponding to the **and_or_inv** *entity shown on page 105.*

An important point illustrated by this example is that the ports of the entity are also visible to processes inside the architecture body and are used in the same way as signals. This corresponds to our view of ports as external pins of a circuit: from the internal point of view, a pin is just a wire with an external connection. So it makes sense for VHDL to treat ports like signals inside an architecture of the entity.

5.3 Behavioral Descriptions

At the most fundamental level, the behavior of a module is described by signal assignment statements within processes. We can think of a process as the basic unit of behavioral description. A process is executed in response to changes of values of signals and uses the present values of signals it reads to determine new values for other signals. A signal assignment is a sequential statement and thus can only appear within a process. In this section, we look in detail at the interaction between signals and processes.

Signal Assignment

In all of the examples we have looked at so far, we have used a simple form of signal assignment statement. Each assignment just provides a new value for a signal. The value is determined by evaluating an expression, the result of which must match the type of the signal. What we have not yet addressed is the issue of timing: when does the signal take on the new value? This is fundamental to modeling hardware, in which events occur over time. First, let us look at the syntax for a basic signal assignment statement in a process:

signal_assignment_statement ⇐
    [label :] name <= [delay_mechanism] waveform ;

waveform ⇐ (*value*_expression [**after** *time*_expression]) { , ₒₒₒ }

The optional label allows us to identify the statement. The syntax rules tell us that we can specify a delay mechanism, which we come to soon, and one or more waveform elements, each consisting of a new value and an optional delay time. It is these delay times in a signal assignment that allow us to specify when the new value should be applied. For example, consider the following assignment:

    y <= **not** or_a_b **after** 5 ns;

This specifies that the signal y is to take on the new value at a time 5 ns later than that at which the statement executes. The delay can be read in one of two ways, depending on whether the model is being used purely for its descriptive value or for simulation. In the first case, the delay can be considered in an abstract sense as a specification of the module's propagation delay: whenever the input changes, the output is updated 5 ns later. In the second case, it can be considered in an operational sense, with reference to a host machine simulating operation of the module by executing the model. Thus if the above assignment is executed at time 250 ns, and or_a_b has the value '1' at that time, then the signal y will take on the value '0' at time 255 ns. Note that the statement itself executes in zero modeled time.

    The time dimension referred to when the model is executed is *simulation time*, that is, the time in which the circuit being modeled is deemed to operate. This is distinct from real execution time on the host machine running a simulation. We measure simulation time starting from zero at the start of execution and increasing in discrete steps as events occur in the model. Not surprisingly, this technique is called *discrete event simulation*. A discrete event simulator must have a simulation time clock, and when a signal assignment statement is executed, the delay specified is added to the current simulation time to determine when the new value is to be applied to the signal. We say that the signal assignment schedules a *transaction* for the signal, where the transaction consists of the new value and the simulation time at which it is to be applied. When simulation time advances to the time at which a transaction is scheduled, the signal is updated with the new value. We say that the signal is *active* during that simulation cycle. If the new value is different from the old value it replaces on a signal, we say an *event* occurs on the signal. The importance of this distinction is that processes respond to events on signals, not to transactions.

The syntax rules for signal assignments show that we can schedule a number of transactions for a signal, to be applied after different delays. For example, a clock driver process might execute the following assignment to generate the next two edges of a clock signal (assuming T_pw is a constant that represents the clock pulse width):

clk <= '1' **after** T_pw, '0' **after** 2*T_pw;

If this statement is executed at simulation time 50 ns and T_pw has the value 10 ns, one transaction is scheduled for time 60 ns to set clk to '1', and a second transaction is scheduled for time 70 ns to set clk to '0'. If we assume that clk has the value '0' when the assignment is executed, both transactions produce events on clk.

This signal assignment statement shows that when more than one transaction is included, the delays are all measured from the current time, not the time in the previous element. Furthermore, the transactions in the list must have strictly increasing delays, so that the list reads in the order that the values will be applied to the signal.

EXAMPLE

We can write a process declaration for a clock generator using the above signal assignment statement to generate a symmetrical clock signal with pulse width T_pw. The difficulty is to get the process to execute regularly every clock cycle. One way to do this is by making it resume whenever the clock changes and scheduling the next two transitions when it changes to '0'. This approach is shown in Figure 5-3.

FIGURE 5-3

```
clock_gen : process (clk) is
begin
    if clk = '0' then
        clk <= '1' after T_pw, '0' after 2*T_pw;
    end if;
end process clock_gen;
```

A process that generates a symmetric clock waveform.

Since a process is the basic unit of a behavioral description, it makes intuitive sense to be allowed to include more than one signal assignment statement for a given signal within a single process. We can think of this as describing the different ways in which a signal's value can be generated by the process at different times.

EXAMPLE

We can write a process that models a two-input multiplexer as shown in Figure 5-4. The value of the **sel** port is used to select which signal assignment to execute to determine the output value.

FIGURE 5-4

```
mux : process (a, b, sel) is
begin
    case sel is
        when '0' =>
            z <= a after prop_delay;
        when '1' =>
            z <= b after prop_delay;
    end case;
end process mux;
```

A process that models a two-input multiplexer.

We say that a process defines a *driver* for a signal if and only if it contains at least one signal assignment statement for the signal. So this example defines a driver for the signal **z**. If a process contains signal assignment statements for several signals, it defines drivers for each of those signals. A driver is a *source* for a signal in that it provides values to be applied to the signal. An important rule to remember is that for normal signals, there may only be one source. This means that we cannot write two different processes each containing signal assignment statements for the one signal. If we want to model such things as buses or wired-or signals, we must use a special kind of signal called a *resolved signal*, which we will discuss in Chapter 8.

VHDL-87

Signal assignment statements may not be labeled in VHDL-87.

Signal Attributes

In Chapter 2 we introduced the idea of attributes of types, which give information about allowed values for the types. Then, in Chapter 4, we saw how we could use attributes of array objects to get information about their index ranges. We can also refer to attributes of signals to find information about their history of transactions and events. Given a signal S, and a value T of type time, VHDL defines the following attributes:

S'delayed(T)	A signal that takes on the same values as S but is delayed by time T.
S'stable(T)	A Boolean signal that is true if there has been no event on S in the time interval T up to the current time, otherwise false.

S'quiet(T)	A Boolean signal that is true if there has been no transaction on S in the time interval T up to the current time, otherwise false.
S'transaction	A signal of type **bit** that changes value from '0' to '1' or vice versa each time there is a transaction on S.
S'event	True if there is an event on S in the current simulation cycle, false otherwise.
S'active	True if there is a transaction on S in the current simulation cycle, false otherwise.
S'last_event	The time interval since the last event on S.
S'last_active	The time interval since the last transaction on S.
S'last_value	The value of S just before the last event on S.

The first three attributes take an optional time parameter. If we omit the parameter, the value 0 fs is assumed. These attributes are often used in checking the timing behavior within a model. For example, we can verify that a signal **d** meets a minimum setup time requirement of **Tsu** before a rising edge on a clock **clk** of type **std_ulogic** as follows:

```
if clk'event and (clk = '1' or clk = 'H')
        and (clk'last_value = '0' or clk'last_value = 'L') then
    assert d'last_event >= Tsu
        report "Timing error: d changed within setup time of clk";
end if;
```

Similarly, we might check that the pulse width of a clock signal input to a module doesn't exceed a maximum frequency by testing its pulse width:

```
assert (not clk'event) or clk'delayed'last_event >= Tpw_clk
    report "Clock frequency too high";
```

Note that we test the time since the last event on a delayed version of the clock signal. When there is currently an event on a signal, the 'last_event attribute returns the value 0 fs. In this case, we determine the time since the previous event by applying the 'last_event attribute to the signal delayed by 0 fs. We can think of this as being an infinitesimal delay. We will return to this idea later in this chapter, in our discussion of delta delays.

EXAMPLE

We can use a similar test for the rising edge of a clock signal to model an edge-triggered module, such as a flipflop. The flipflop should load the value of its D input on a rising edge of **clk**, but asynchronously clear the outputs whenever **clr** is '1'. The entity declaration and a behavioral architecture body are shown in Figure 5-5.

If the flipflop did not have the asynchronous clear input, the model could have used a simple wait statement such as

```
wait until clk = '1';
```

to trigger on a rising edge. However, with the clear input present, the process must be sensitive to changes on both clk and clr at any time. Hence it uses the 'event attribute to distinguish between clk changing to '1' and clr going back to '0' while clk is stable at '1'.

FIGURE 5-5

```
entity edge_triggered_Dff is
    port ( D : in bit; clk : in bit; clr : in bit;
          Q : out bit );
end entity edge_triggered_Dff;

–––––––––––––––––––––––––––––––––––––––––––––––

architecture behavioral of edge_triggered_Dff is
begin
    state_change : process (clk, clr) is
    begin
        if clr = '1' then
            Q <= '0' after 2 ns;
        elsif clk'event and clk = '1' then
            Q <= D after 2 ns;
        end if;
    end process state_change;
end architecture behavioral;
```

An entity and architecture body for an edge-triggered flipflop, using the 'event *attribute to check for changes on the* clk *signal.*

VHDL-87

In VHDL-87, the 'last_value attribute for a composite signal returns the aggregate of last values for each of the scalar elements of the signal. For example, suppose a bit-vector signal **s** initially has the value B"00" and changes to B"01" and then B"11" in successive events. After the last event, the result of s'last_value is B"00" in VHDL-87. In VHDL-93 it is B"01", since that is the last value of the entire composite signal.

Wait Statements

Now that we have seen how to change the values of signals over time, the next step in behavioral modeling is to specify when processes respond to changes in signal values. This is done using *wait statements*. A wait statement is a sequential statement with the following syntax rule:

```
wait_statement ⇐
    〚 label : 〛 wait 〚 on signal_name { , ... } 〛
                    〚 until boolean_expression 〛
                    〚 for time_expression 〛 ;
```

The optional label allows us to identify the statement. The purpose of the wait statement is to cause the process that executes the statement to suspend execution. The *sensitivity* clause, *condition* clause and *timeout* clause specify when the process is subsequently to resume execution. We can include any combination of these clauses, or we may omit all three. Let us go through each clause and describe what it specifies.

The sensitivity clause, starting with the word **on**, allows us to specify a list of signals to which the process responds. If we just include a sensitivity clause in a wait statement, the process will resume whenever any one of the listed signals changes value, that is, whenever an event occurs on any of the signals. This style of wait statement is useful in a process that models a block of combinatorial logic, since any change on the inputs may result in new output values; for example:

```
half_add : process is
begin
    sum <= a xor b after T_pd;
    carry <= a and b after T_pd;
    wait on a, b;
end process half_add;
```

The process starts execution by generating values for **sum** and **carry** based on the initial values of **a** and **b**, then suspends on the wait statement until either **a** or **b** (or both) change values. When that happens, the process resumes and starts execution from the top.

This form of process is so common in modeling digital systems that VHDL provides the shorthand notation that we have seen in many examples in preceding chapters. A process with a sensitivity list in its heading is exactly equivalent to a process with a wait statement at the end, containing a sensitivity clause naming the signals in the sensitivity list. So the **half_add** process above could be rewritten as

```
half_add : process (a, b) is
begin
    sum <= a xor b after T_pd;
    carry <= a and b after T_pd;
end process half_add;
```

EXAMPLE

Let us return to the model of a two-input multiplexer shown in Figure 5-4. The process in that model is sensitive to all three input signals. This means that it will resume on changes on either data input, even though only one of them is selected at any time. If we are concerned about this slight lack of efficiency in simulation, we can write the process differently, using wait statements to be more selective about the signals to which the process is sensitive each time it suspends. The revised model is shown in Figure 5-6. In this model, when input **a** is selected, the process only waits for changes on the select input and on **a**. Any changes on **b** are ignored. Similarly, if **b** is selected, the process waits for changes on **sel** and on **b**, ignoring changes on **a**.

FIGURE 5-6

```
entity mux2 is
    port ( a, b, sel : in bit;
            z : out bit );
end entity mux2;

- - - - - - - - - - - - - - - - - - - - - - - - - - - - - - - - - - - - - - - -

architecture behavioral of mux2 is
    constant prop_delay : time := 2 ns;
begin
    slick_mux : process is
    begin
        case sel is
            when '0' =>
                z <= a after prop_delay;
                wait on sel, a;
            when '1' =>
                z <= b after prop_delay;
                wait on sel, b;
        end case;
    end process slick_mux;
end architecture behavioral;
```

An entity and architecture body for a multiplexer that avoids being resumed in response to changes on the input signal that is not currently selected.

The condition clause in a wait statement, starting with the word **until**, allows us to specify a condition that must be true for the process to resume. For example, the wait statement

wait until clk = '1';

causes the executing process to suspend until the value of the signal clk changes to '1'. The condition expression is tested while the process is suspended to determine whether to resume the process. A consequence of this is that even if the condition is true when the wait statement is executed, the process will still suspend until the appropriate signals change and cause the condition to be true again. If the wait statement doesn't include a sensitivity clause, the condition is tested whenever an event occurs on any of the signals mentioned in the condition.

EXAMPLE

The clock generator process from the example on page 110 can be rewritten using a wait statement with a condition clause, as shown in Figure 5-7. Each time the process executes the wait statement, clk has the value '0'. However, the process still suspends, and the condition is tested each time there is an event on clk. When clk changes to '1', nothing happens, but when it changes to '0' again, the process resumes and schedules transactions for the next cycle.

FIGURE 5-7

```
clock_gen : process is
begin
    clk <= '1' after T_pw, '0' after 2*T_pw;
    wait until clk = '0';
end process clock_gen;
```

The revised clock generator process.

If a wait statement includes a sensitivity clause as well as a condition clause, the condition is only tested when an event occurs on any of the signals in the sensitivity clause. For example, if a process suspends on the following wait statement:

wait on clk **until** reset = '0';

the condition is tested on each change in the value of **clk**, regardless of any changes on **reset**.

The timeout clause in a wait statement, starting with the word **for**, allows us to specify a maximum interval of simulation time for which the process should be suspended. If we also include a sensitivity or condition clause, these may cause the process to be resumed earlier. For example, the wait statement

wait until trigger = '1' **for** 1 ms;

causes the executing process to suspend until **trigger** changes to '1', or until 1 ms of simulation time has elapsed, whichever comes first. If we just include a timeout clause by itself in a wait statement, the process will suspend for the time given.

EXAMPLE

We can rewrite the clock generator process from the example on page 110 yet again, this time using a wait statement with a timeout clause, as shown in Figure 5-8. In this case we specify the clock period as the timeout, after which the process is to be resumed.

FIGURE 5-8

```
clock_gen : process is
begin
    clk <= '1' after T_pw, '0' after 2*T_pw;
    wait for 2*T_pw;
end process clock_gen;
```

A third version of the clock generator process.

If we refer back to the syntax rule for a wait statement shown on page 113, we note that it is legal to write

wait;

This form causes the executing process to suspend for the remainder of the simulation. Although this may at first seem a strange thing to want to do, in practice it is quite useful. One place where it is used is in a process whose purpose is to generate stimuli for a simulation. Such a process should generate a sequence of transactions on signals connected to other parts of a model and then stop. For example, the process

```
test_gen : process is
begin
    test0 <= '0' after 10 ns, '1' after 20 ns, '0' after 30 ns, '1' after 40 ns;
    test1 <= '0' after 10 ns, '1' after 30 ns;
    wait;
end process test_gen;
```

generates all four possible combinations of values on the signals test0 and test1. If the final wait statement were omitted, the process would cycle forever, repeating the signal assignment statements without suspending, and the simulation would make no progress.

VHDL-87

Wait statements may not be labeled in VHDL-87.

Delta Delays

Let us now return to the topic of delays in signal assignments. In many of the example signal assigments in previous chapters, we omitted the delay part of waveform elements. This is equivalent to specifying a delay of 0 fs. The value is to be applied to the signal at the current simulation time. However, it is important to note that the signal value does not change as soon as the signal assignment statement is executed. Rather, the assignment schedules a transaction for the signal, which is applied after the process suspends. Thus the process does not see the effect of the assignment until the next time it resumes, even if this is at the same simulation time. For this reason, a delay of 0 fs in a signal assignment is called a *delta delay*.

To understand why delta delays work in this way, it is necessary to review the simulation cycle, introduced in Chapter 1 on page 15. Recall that the simulation cycle consists of two phases: a signal update phase followed by a process execution phase. In the signal update phase, simulation time is advanced to the time of the earliest scheduled transaction, and the values in all transactions scheduled for this time are applied to their corresponding signals. This may cause events to occur on some signals. In the process execution phase, all processes that respond to these events are resumed and execute until they suspend again on wait statements. The simulator then repeats the simulation cycle.

Let us now consider what happens when a process executes a signal assignment statement with delta delay, for example:

```
data <= X"00";
```

Suppose this is executed at simulation time *t* during the process execution phase of the current simulation cycle. The effect of the assignment is to schedule a transaction to put the value X"00" on **data** at time *t*. The transaction is not applied immediately, since the simulator is in the process execution phase. Hence the process continues executing, with **data** unchanged. When all processes have suspended, the simulator starts the next simulation cycle and updates the simulation time. Since the earliest transaction is now at time *t*, simulation time remains unchanged. The simulator now applies the value X"00" in the scheduled transaction to **data**, then resumes any processes that respond to the new value.

Writing a model with delta delays is useful when we are working at a high level of abstraction and are not yet concerned with detailed timing. If all we are interested in is describing the order in which operations take place, delta delays provide a means of ignoring the complications of timing. We have seen this in many of the examples in previous chapters. However, we should note a common pitfall encountered by most beginner VHDL designers when using delta delays: they forget that the process does not see the effect of the assignment immediately. For example, we might write a process that includes the following statements:

```
s <= '1';
. . .
if s = '1' then . . .
```

and expect the process to execute the if statement assuming **s** has the value '1'. We would then spend fruitless hours debugging our model until we remembered that **s** still has its old value until the next simulation cycle, after the process has suspended.

EXAMPLE

Figure 5-9 is an outline of an abstract model of a computer system. The CPU and memory are connected with address and data signals. They synchronize their operation with the **mem_read** and **mem_write** control signals and the **mem_ready** status signal. No delays are specified in the signal assignment statements, so synchronization occurs over a number of delta delay cycles, as shown in Figure 5-10.

FIGURE 5-9

```
architecture abstract of computer_system is
    subtype word is bit_vector(31 downto 0);
    signal address : natural;
    signal read_data, write_data : word;
    signal mem_read, mem_write : bit := '0';
    signal mem_ready : bit := '0';
begin
```

```
cpu : process is
    variable instr_reg : word;
    variable PC : natural;
    . . .        - - other declarations
begin
    loop
        address <= PC;
        mem_read <= '1';
        wait until mem_ready = '1';
        instr_reg := read_data;
        mem_read <= '0';
        wait until mem_ready = '0';
        PC := PC + 4;
        . . .          - - execute the instruction
    end loop;
end process cpu;

memory : process is
    type memory_array is array (0 to 2**14 - 1) of word;
    variable store : memory_array := (
        . . .
        );
begin
    wait until mem_read = '1' or mem_write = '1';
    if mem_read = '1' then
        read_data <= store( address / 4 );
        mem_ready <= '1';
        wait until mem_read = '0';
        mem_ready <= '0';
    else
        . . .          - - perform write access
    end if;
end process memory;
end architecture abstract;
```

An outline of an abstract model for a computer system, consisting of a CPU and a memory. The processes use delta delays to synchronize communication, rather than modeling timing of bus transactions in detail.

When the simulation starts, the CPU process begins executing its statements and the memory suspends. The CPU schedules transactions to assign the next instruction address to the **address** signal and the value '1' to the **mem_read** signal, then suspends. In the next simulation cycle, these signals are updated and the memory process resumes, since it is waiting for an event on **mem_read**. The memory process schedules the data on the **read_data** signal and the value '1' on **mem_ready**, then suspends. In the third cycle, these signals are updated and the CPU process resumes. It schedules the value '0' on **mem_read** and suspends. Then, in the fourth cycle, **mem_read** is updated and the memory process is resumed, scheduling the value '0' on **mem_ready** to complete the handshake. Finally, on the fifth cycle, **mem_ready** is updated and the CPU process resumes and executes the fetched instruction.

FIGURE 5-10

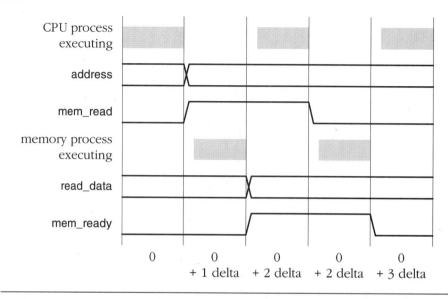

Synchronization over successive delta cycles in a simulation of a read operation between the CPU and memory shown in Figure 5-9.

Transport and Inertial Delay Mechanisms

So far in our discussion of signal assignments, we have implicitly assumed that there were no pending transactions scheduled for a signal when a signal assignment statement was executed. In many models, particularly at higher levels of abstraction, this will be the case. If, on the other hand, there are pending transactions, the new transactions are merged with them in a way that depends on the *delay mechanism* used in the signal assignment statement. This is an optional part of the signal assignment syntax shown on page 109. The syntax rule for the delay mechanism is

delay_mechanism ⇐ **transport** ▯ ⟦ **reject** *time*_expression ⟧ **inertial**

A signal assignment with the delay mechanism part omitted is equivalent to specifying **inertial**. We look at the *transport* delay mechanism first, since it is simpler, and then return to the *inertial* delay mechanism.

We use the transport delay mechanism when we are modeling an ideal device with infinite frequency response, in which any input pulse, no matter how short, produces an output pulse. An example of such a device is an ideal transmission line, which transmits all input changes delayed by some amount. A process to model a transmission line with delay 500 ps is

```
transmission_line : process (line_in) is
begin
    line_out <= transport line_in after 500 ps;
end process transmission_line;
```

In this model the output follows any changes in the input, but delayed by 500 ps. If the input changes twice or more within a period shorter than 500 ps, the scheduled transactions are simply queued by the driver until the simulation time at which they are to be applied, as shown in Figure 5-11.

FIGURE 5-11

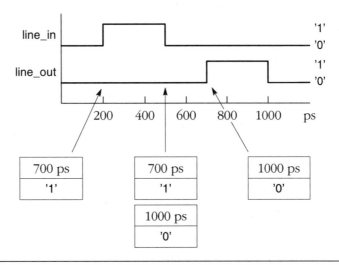

Transactions queued by a driver using transport delay. At time 200 ps the input changes, and a transaction is scheduled for 700 ps. At time 500 ps, the input changes again, and another transaction is scheduled for 1000 ps. This is queued by the driver behind the earlier transaction. When simulation time reaches 700 ps, the first transaction is applied, and the second transaction remains queued. Finally, simulation time reaches 1000 ps, and the final transaction is applied, leaving the driver queue empty.

In this example, each new transaction that is generated by a signal assignment statement is scheduled for a simulation time that is later than the pending transactions queued by the driver. The situation gets a little more complex when variable delays are used, since we can schedule a transaction for an earlier time than a pending transaction. The semantics of the transport delay mechanism specify that if there are pending transactions on a driver that are scheduled for a time later than or equal to a new transaction, those later transactions are deleted.

EXAMPLE

Figure 5-12 is a process that describes the behavior of an asymmetric delay element, with different delay times for rising and falling transitions. The delay for rising transitions is 800 ps and for falling transitions 500 ps. If we apply an input pulse of only 200 ps duration, we would expect the output not to change, since the delayed falling transition should "overtake" the delayed rising transition. If we were simply to add each transition to the driver queue when a signal assignment statement is executed, we would not get this behavior. However, the semantics of the transport delay mechanism produce the desired behavior, as Figure 5-13 shows.

FIGURE 5-12

```
asym_delay : process (a) is
    constant Tpd_01 : time := 800 ps;
    constant Tpd_10 : time := 500 ps;
begin
    if a = '1' then
        z <= transport a after Tpd_01;
    else  --a = '0'
        z <= transport a after Tpd_10;
    end if;
end process asym_delay;
```

A process that describes a delay element with asymmetric delays for rising and falling transitions.

FIGURE 5-13

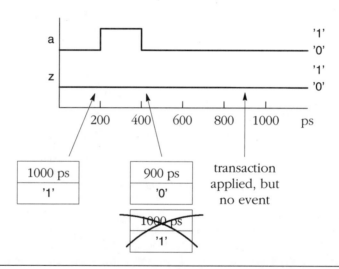

*Transactions in a driver using asymmetric transport delay. At time 200 ps the input changes, and a trans-
action is scheduled for 1000 ps. At time 400 ps, the input changes again, and another transaction is
scheduled for 900 ps. Since this is earlier than the pending transaction at 1000 ps, the pending transaction
is deleted. When simulation time reaches 900 ps, the remaining transaction is applied, but since the value
is '0', no event occurs on the signal.*

Most real electronic circuits don't have infinite frequency response, so it is not appropriate to model them using transport delay. In real devices, changing the values
of internal nodes and outputs involves moving electronic charge around in the presence of capacitance, inductance and resistance. This gives the device some inertia; it
tends to stay in the same state unless we force it by applying inputs for a sufficiently
long duration. This is why VHDL includes the inertial delay mechanism, to allow us
to model devices that reject input pulses too short to overcome their inertia. Inertial
delay is the mechanism used by default in a signal assignment, or we can specify it
explicitly by including the word **inertial**.

To explain how inertial delay works, let us first consider a model in which all the signal assignments for a given signal use the same delay value, say 3 ns, as in the following inverter model:

```
inv : process (a) is
begin
       y <= inertial not a after 3 ns;
end process inv;
```

So long as input events occur more than 3 ns apart, this model does not present any problems. Each time a signal assignment is executed, there are no pending transactions, so a new transaction is scheduled, and the output changes value 3 ns later. However, if an input changes less than 3 ns after the previous change, this represents a pulse less than the propagation delay of the device, so it should be rejected. This behavior is shown at the top of Figure 5-14. In a simple model such as this, we can interpret inertial delay as saying if a signal assignment would produce an output pulse shorter than the propagation delay, then the output pulse does not happen.

FIGURE 5-14

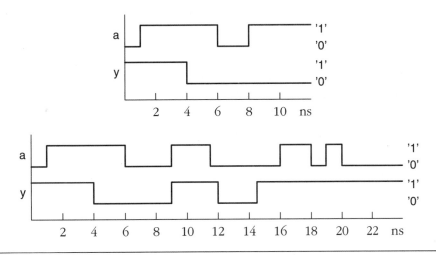

Results of signal assignments using the inertial delay mechanism. In the top waveform, an inertial delay of 3 ns is specified. The input change at time 1 ns is reflected in the output at time 4 ns. The pulse from 6 to 8 ns is less than the propagation delay, so it doesn't affect the output. In the bottom waveform, an inertial delay of 3 ns and a pulse rejection limit of 2 ns are specified. The input changes at 1, 6, 9 and 11.5 ns are all reflected in the output, since they occur greater than 2 ns apart. However, the subsequent input pulses are less than or equal to the pulse rejection limit in length, and so do not affect the output.

Next, let us extend this model by specifying a pulse rejection limit, after the word **reject** in the signal assignment:

```
inv : process (a) is
begin
       y <= reject 2 ns inertial not a after 3 ns;
end process inv;
```

We can interpret this as saying if a signal assignment would produce an output pulse shorter than (or equal to) the pulse rejection limit, the output pulse does not happen. In this simple model, so long as input changes occur more than 2 ns apart, they produce output changes 3 ns later, as shown at the bottom of Figure 5-14. Note that the pulse rejection limit specified must be between 0 fs and the delay specified in the signal assignment. Omitting a pulse rejection limit is the same as specifying a limit equal to the delay, and specifying a limit of 0 fs is the same as specifying transport delay.

Now let us look at the full story of inertial delay, allowing for varying the delay time and pulse rejection limit in different signal assignments applied to the same signal. As with transport delay, the situation becomes more complex, and it is best to describe it in terms of deleting transactions from the driver. Those who are unlikely to be writing models that deal with timing at this level of detail may wish to move on to the next section.

An inertially delayed signal assignment involves examining the pending transactions on a driver when adding a new transaction. Suppose a signal assignment schedules a new transaction for time t_{new}, with a pulse rejection limit of t_r. First, any pending transactions scheduled for a time later than or equal to t_{new} are deleted, just as they are when transport delay is used. Then the new transaction is added to the driver. Second, any pending transactions scheduled in the interval $t_{new} - t_r$ to t_{new} are examined. If there is a run of consecutive transactions immediately preceding the new transaction with the same value as the new transaction, they are kept in the driver. All other transactions in the interval are deleted.

An example will make this clearer. Suppose a driver for signal **s** contains pending transactions as shown at the top of Figure 5-15, and the process containing the driver executes the following signal assignment statement at time 10 ns:

s <= **reject** 5 ns **inertial** '1' **after** 8 ns;

The pending transactions after this assignment are shown at the bottom of Figure 5-15.

FIGURE 5-15

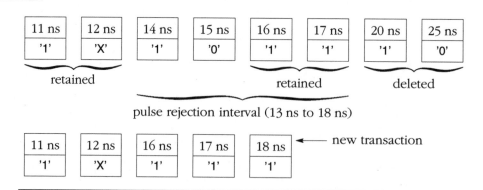

Transactions before (top) and after (bottom) an inertial delay signal assignment. The transactions at 20 and 25 ns are deleted because they are scheduled for later than the new transaction. Those at 11 and 12 ns are retained because they fall before the pulse rejection interval. The transactions at 16 and 17 ns fall within the rejection interval, but they form a run leading up to the new transaction, with the same value as the new transaction; hence they are also retained. The other transactions in the rejection interval are deleted.

One final point to note about specifying the delay mechanism in a signal assignment statement is that if a number of waveform elements are included, the specified mechanism only applies to the first element. All the subsequent elements schedule transactions using transport delay. Since the delays for multiple waveform elements must be in ascending order, this means that all of the transactions after the first are just added to the driver transaction queue in the order written.

EXAMPLE

A detailed model of a two-input and gate is shown in Figure 5-16. When a change on either of the input signals results in a change scheduled for the output, the **delay** process determines the propagation delay to be used. On a rising output transition, spikes of less than 400 ps are rejected, and on a falling or unknown transition, spikes of less than 300 ps are rejected. Note that the result of the **and** operator, when applied to standard-logic values, is always 'U', 'X', '0' or '1'. Hence the **delay** process need not compare **result** with 'H' or 'L' when testing for rising or falling transitions.

FIGURE 5-16

```
library ieee;  use ieee.std_logic_1164.all;
entity and2 is
    port ( a, b : in std_ulogic;  y : out std_ulogic );
end entity and2;

--------------------------------------------------

architecture detailed_delay of and2 is
    signal result : std_ulogic;
begin
    gate : process (a, b) is
    begin
        result <= a and b;
    end process gate;
    delay : process (result) is
    begin
        if result = '1' then
            y <= reject 400 ps inertial '1' after 1.5 ns;
        elsif result = '0' then
            y <= reject 300 ps inertial '0' after 1.2 ns;
        else
            y <= reject 300 ps inertial 'X' after 500 ps;
        end if;
    end process delay;
end architecture detailed_delay;
```

An entity and architecture body for a two-input and gate. The process gate *implements the logical function of the entity, and the process* delay *implements its detailed timing characteristics using inertially delayed signal assignments. A delay of 1.5 ns is used for rising transitions, 1.2 ns for falling transitions.*

VHDL-87

VHDL-87 does not allow specification of the pulse rejection limit in a delay mechanism. The syntax rule in VHDL-87 is

delay_mechanism ⇐ **transport**

If the delay mechanism is omitted, inertial delay is used, with a pulse rejection limit equal to the delay specified in the waveform element.

Process Statements

We have been using processes quite extensively in examples in this and previous chapters, so we have seen most of the details of how they are written and used. To summarize, let us now look at the formal syntax for a process statement and review process operation. The syntax rule is

process_statement ⇐
 [*process*_label :]
 process [(*signal*_name { , ... })] [**is**]
  { process_declarative_item }
 begin
  { sequential_statement }
 end process [*process*_label] ;

Recall that a process statement is a concurrent statement that can be included in an architecture body to implement all or part of the behavior of a module. The process label identifies the process. While it is optional, it is a good idea to include a label on each process. A label makes it easier to debug a simulation of a system, since most simulators provide a way of identifying a process by its label. Most simulators also generate a default name for a process if we omit the label in the process statement. Having identified a process, we can examine the contents of its variables or set breakpoints at statements within the process.

The declarative items in a process statement may include constant, type and variable declarations, as well as other declarations that we will come to later. Note that ordinary variables may only be declared within process statements, not outside of them. The variables are used to represent the state of the process, as we have seen in the examples. The sequential statements that form the process body may include any of those that we introduced in Chapter 3, plus signal assignment and wait statements. When a process is activated during simulation, it starts executing from the first sequential statement and continues until it reaches the last. It then starts again from the first. This would be an infinite loop, with no progress being made in the simulation, if it were not for the inclusion of wait statements, which suspend process execution until some relevant event occurs. Wait statements are the only statements that take more than zero simulation time to execute. It is only through the execution of wait statements that simulation time advances.

A process may include a sensitivity list in parentheses after the keyword **process**. The sensitivity list identifies a set of signals that the process monitors for events. If the sensitivity list is omitted, the process should include one or more wait statements. On the other hand, if the sensitivity list is included, then the process body cannot include any wait statements. Instead, there is an implicit wait statement, just before the **end process** keywords, that includes the signals listed in the sensitivity list as signals in an **on** clause.

VHDL-87

The keyword **is** may not be included in the header of a process statement in VHDL-87.

Concurrent Signal Assignment Statements

The form of process statement that we have been using is the basis for all behavioral modeling in VHDL, but for simple cases, it can be a little cumbersome and verbose. For this reason, VHDL provides us with some useful shorthand notations for *functional* modeling, that is, behavioral modeling in which the operation to be described is a simple combinatorial transformation of inputs to an output. We look at the basic form of two of these statements, *concurrent signal assignment* statements, which are concurrent statements that are essentially signal assignments. Unlike ordinary signal assignments, concurrent signal assignment statements can be included in the statement part of an architecture body. The syntax rule is

concurrent_signal_assignment_statement ⇐
    ⟦ label : ⟧ conditional_signal_assignment
    ❙ ⟦ label : ⟧ selected_signal_assignment

which tells us that the two forms are called a *conditional signal assignment* and a *selected signal assignment*. Each of them may include a label, which serves exactly the same purpose as a label on a process statement: it allows the statement to be identified by name during simulation or synthesis.

Conditional Signal Assignment Statements

The conditional signal assignment statement is a shorthand for a collection of ordinary signal assignments contained in an if statement, which is in turn contained in a process statement. The simplified syntax rule for a conditional signal assignment is

conditional_signal_assignment ⇐
    name <= ⟦ delay_mechanism ⟧
               ❴ waveform **when** *boolean*_expression **else** ❵
              waveform ⟦ **when** *boolean*_expression ⟧ ;

The conditional signal assignment allows us to specify which of a number of waveforms should be assigned to a signal depending on the values of some conditions. Let us look at some examples and show how each conditional signal assignment can be transformed into an equivalent process statement. First, the top statement in Figure 5-17 is a functional description of a multiplexer, with four data inputs (d0, d1, d2 and d3), two select inputs (sel0 and sel1) and a data output (z). All of these signals are of type bit. This statement has exactly the same meaning as the process statement shown at the bottom of Figure 5-17.

FIGURE 5-17

```
zmux : z <= d0 when sel1 = '0' and sel0 = '0' else
              d1 when sel1 = '0' and sel0 = '1' else
              d2 when sel1 = '1' and sel0 = '0' else
              d3 when sel1 = '1' and sel0 = '1';

----------------------------------------------

zmux : process is
begin
    if sel1 = '0' and sel0 = '0' then
        z <= d0;
    elsif sel1 = '0' and sel0 = '1' then
        z <= d1;
    elsif sel1 = '1' and sel0 = '0' then
        z <= d2;
    elsif sel1 = '1' and sel0 = '1' then
        z <= d3;
    end if;
    wait on d0, d1, d2, d3, sel0, sel1;
end process zmux;
```

Top: a functional model of a multiplexer, using a conditional signal assignment statement. Bottom: the equivalent process statement.

The advantage of the conditional signal assignment form over the equivalent process is clearly evident from this example. The simple combinatorial transformation is obvious to the reader, uncluttered by the details of the process mechanism. This is not to say that processes are a bad thing, rather that in simple cases, we would rather hide that detail to make the model clearer. Looking at the equivalent process shows us something important about the conditional signal assignment statement, namely, that it is sensitive to all of the signals mentioned in the waveforms and the conditions. So whenever any of these change value, the conditional assignment is reevaluated and a new transaction scheduled on the driver for the target signal.

If we look more closely at the multiplexer model, we note that the last condition is redundant, since the signals sel0 and sel1 are of type bit. If none of the previous conditions are true, the signal should always be assigned the last waveform. So we can rewrite the example as shown in Figure 5-18.

FIGURE 5-18

```
zmux : z <= d0 when sel1 = '0' and sel0 = '0' else
              d1 when sel1 = '0' and sel0 = '1' else
              d2 when sel1 = '1' and sel0 = '0' else
              d3;

----------------------------------------------
```

```
zmux : process is
begin
    if sel1 = '0' and sel0 = '0' then
        z <= d0;
    elsif sel1 = '0' and sel0 = '1' then
        z <= d1;
    elsif sel1 = '1' and sel0 = '0' then
        z <= d2;
    else
        z <= d3;
    end if;
    wait on d0, d1, d2, d3, sel0, sel1;
end process zmux;
```

A revised functional model for the multiplexer, with its equivalent process statement.

A very common case in function modeling is to write a conditional signal assignment with no conditions, as in the following example:

```
PC_incr : next_PC <= PC + 4 after 5 ns;
```

At first sight this appears to be an ordinary sequential signal assignment statement, which by rights ought to be inside a process body. However, if we look at the syntax rule for a concurrent signal assignment, we note that this can in fact be recognized as such if all of the optional parts except the label are omitted. In this case, the equivalent process statement is

```
PC_incr : process is
begin
    next_PC <= PC + 4 after 5 ns;
    wait on PC;
end process PC_incr;
```

Another case that sometimes arises when writing functional models is the need for a process that schedules an initial set of transactions and then does nothing more for the remainder of the simulation. An example is the generation of a reset signal. One way of doing this is as follows:

```
reset_gen : reset <= '1', '0' after 200 ns when extended_reset else
                     '1', '0' after 50 ns;
```

The thing to note here is that there are no signals named in any of the waveforms or the conditions (assuming that extended_reset is a constant). This means that the statement is executed once when simulation starts, schedules two transactions on reset and remains quiescent thereafter. The equivalent process is

```
reset_gen : process is
begin
    if extended_reset then
        reset <= '1', '0' after 200 ns;
```

```
    else
        reset <= '1', '0' after 50 ns;
    end if;
    wait;
end process reset_gen;
```

Since there are no signals involved, the wait statement has no sensitivity clause. Thus after the if statement has executed, the process suspends forever.

If we include a delay mechanism specification in a conditional signal assignment statement, it is used whichever waveform is chosen. So we might rewrite the model for the asymmetric delay element shown in Figure 5-12 as

```
asym_delay : z <= transport a after Tpd_01 when a = '1' else
                  a after Tpd_10;
```

One problem with conditional signal assignments, as we have described them so far, is that they always assign a new value to a signal. Sometimes we may not want to change the value of a signal, or more specifically, we may not want to schedule any new transactions on the signal. We can use the keyword **unaffected** instead of a normal waveform for these cases, as shown at the top of Figure 5-19.

FIGURE 5-19

```
scheduler :
    request <= first_priority_request after scheduling_delay
                    when priority_waiting and server_status = ready else
              first_normal_request after scheduling_delay
                    when not priority_waiting and server_status = ready else
              unaffected
                    when server_status = busy else
              reset_request after scheduling_delay;
--------------------------------------------------
scheduler : process is
begin
    if priority_waiting and server_status = ready then
        request <= first_priority_request after scheduling_delay;
    elsif not priority_waiting and server_status = ready then
        request <= first_normal_request after scheduling_delay;
    elsif server_status = busy then
        null;
    else
        request <= reset_request after scheduling_delay;
    end if;
    wait on first_priority_request, priority_waiting, server_status,
                first_normal_request, reset_request;
end process scheduler;
```

*Top: a conditional signal assignment statement showing use of the **unaffected** waveform. Bottom: the equivalent process statement.*

The effect of the **unaffected** waveform is to include a null statement in the equivalent process, causing it to bypass scheduling a transaction when the corresponding condition is true. (Recall that the effect of the null sequential statement is to do nothing.) So the example at the top of Figure 5-19 is equivalent to the process shown at the bottom. Note that we can only use **unaffected** in a concurrent signal assignment, not in a sequential signal assignment.

VHDL-87

In VHDL-87 the syntax rule for a conditional signal assignment statement is

conditional_signal_assignment ⟸
    name <= ⟦ **transport** ⟧
            { waveform **when** *boolean*_expression **else** }
            waveform ;

The delay mechanism is restricted to the keyword **transport**. The final waveform may not be conditional. Furthermore, we may not use the keyword **unaffected**. If the required behavior cannot be expressed with these restrictions, we must write a full process statement instead of a conditional signal assignment statement.

Selected Signal Assignment Statements

The selected signal assignment statement is similar in many ways to the conditional signal assignment statement. It, too, is a shorthand for a number of ordinary signal assignments embedded in a process. But for a selected signal assignment, the equivalent process contains a case statement instead of an if statement. The simplified syntax rule is

selected_signal_assignment ⟸
    with expression **select**
        name <= ⟦ delay_mechanism ⟧
                { waveform **when** choices , }
                waveform **when** choices ;

This statement allows us to choose between a number of waveforms to be assigned to a signal depending on the value of an expression. As an example, let us consider the selected signal assignment shown at the top of Figure 5-20. This has the same meaning as the process statement containing a case statement shown at the bottom of Figure 5-20.

A selected signal assignment statement is sensitive to all of the signals in the selector expression and in the waveforms. This means that the selected signal assignment in Figure 5-20 is sensitive to b and will resume if b changes value, even if the value of alu_function is alu_pass_a.

An important point to note about a selected signal assignment statement is that the case statement in the equivalent process must be legal according to all of the rules that we described in Chapter 3. This means that every possible value for the selector expression must be accounted for in one of the choices, that no value is included in more than one choice and so on.

FIGURE 5-20

```
alu : with alu_function select
        result <=  a + b after Tpd        when alu_add | alu_add_unsigned,
                   a – b after Tpd        when alu_sub | alu_sub_unsigned,
                   a and b after Tpd  when alu_and,
                   a or b after Tpd    when alu_or,
                   a after Tpd            when alu_pass_a;
```

- -

```
alu : process is
begin
    case alu_function is
        when alu_add | alu_add_unsigned  => result <= a + b after Tpd;
        when alu_sub | alu_sub_unsigned => result <= a – b after Tpd;
        when alu_and                            => result <= a and b after Tpd;
        when alu_or                              => result <= a or b after Tpd;
        when alu_pass_a                        => result <= a after Tpd;
    end case;
    wait on alu_function, a, b;
end process alu;
```

Top: a selected signal assignment. Bottom: its equivalent process statement.

Apart from the difference in the equivalent process, the selected signal assignment is similar to the conditional assignment. Thus the special waveform **unaffected** can be used to specify that no assignment take place for some values of the selector expression. Also, if a delay mechanism is specified in the statement, that mechanism is used on each sequential signal assignment within the equivalent process.

EXAMPLE

We can use a selected signal assignment to express a combinatorial logic function in truth-table form. Figure 5-21 shows an entity declaration and an architecture body for a full adder. The selected signal assignment statement has, as its selector expression, a bit vector formed by aggregating the input signals. The choices list all possible values of inputs, and for each, the values for the c_out and s outputs are given.

FIGURE 5-21

```
entity full_adder is
    port ( a, b, c_in : bit;  s, c_out : out bit );
end entity full_adder;
```

- -

```
architecture truth_table of full_adder is
begin
    with bit_vector'(a, b, c_in) select
        (c_out, s) <= bit_vector'("00") when "000",
                          bit_vector'("01") when "001",
```

```
                    bit_vector'("01") when "010",
                    bit_vector'("10") when "011",
                    bit_vector'("01") when "100",
                    bit_vector'("10") when "101",
                    bit_vector'("10") when "110",
                    bit_vector'("11") when "111";
        end architecture truth_table;
```

An entity declaration and functional architecture body for a full adder.

This example illustrates the most common use of aggregate targets in signal assignments. Note that the type qualification is required in the selector expression to specify the type of the aggregate. The type qualification is needed in the output values to distinguish the bit-vector string literals from character string literals.

VHDL-87

In VHDL-87, the delay mechanism is restricted to the keyword **transport**, as discussed on page 126. Furthermore, the keyword **unaffected** may not be used. If the required behavior cannot be expressed without using the keyword **unaffected**, we must write a full process statement instead of a selected signal assignment statement.

Concurrent Assertion Statements

VHDL provides another shorthand process notation, the *concurrent assertion statement*, which can be used in behavioral modeling. As its name implies, a concurrent assertion statement represents a process whose body contains an ordinary sequential assertion statement. The syntax rule is

```
concurrent_assertion_statement ⇐
    [ label : ]
    assert boolean_expression
        [ report expression ] [ severity expression ] ;
```

This syntax appears to be exactly the same as that for a sequential assertion statement, but the difference is that it may appear as a concurrent statement. The optional label on the statement serves the same purpose as that on a process statement: to provide a way of referring to the statement during simulation or synthesis. The process equivalent to a concurrent assertion contains a sequential assertion with the same condition, report clause and severity clause. The sequential assertion is then followed by a wait statement whose sensitivity list includes the signals mentioned in the condition expression. Thus the effect of the concurrent assertion statement is to check that the condition holds true each time any of the signals mentioned in the condition change value. Concurrent assertions provide a very compact and useful way of including timing and correctness checks in a model.

EXAMPLE

We can use concurrent assertion statements to check for correct use of a set/reset flipflop, with two inputs s and r and two outputs q and q_n, all of type bit. The requirement for use is that s and r are not both '1' at the same time. The entity and architecture body are shown in Figure 5-22.

FIGURE 5-22

```
entity S_R_flipflop is
    port ( s, r : in bit;  q, q_n : out bit );
end entity S_R_flipflop;

- - - - - - - - - - - - - - - - - - - - - - - - - - - - - - - - - - - - - - -

architecture functional of S_R_flipflop is
begin
    q <= '1' when s = '1' else
         '0' when r = '1';
    q_n <= '0' when s = '1' else
           '1' when r = '1';
    check : assert not (s = '1' and r = '1')
                report "Incorrect use of S_R_flip_flop: s and r both '1'";
end architecture functional;
```

An entity and architecture body for a set/reset flipflop, including a concurrent assertion statement to check for correct usage.

The first and second concurrent statements implement the functionality of the model. The third checks for correct use and is resumed when either s or r changes value, since these are the signals mentioned in the Boolean condition. If both of the signals are '1', an assertion violation is reported. The equivalent process for the concurrent assertion is

```
check : process is
begin
    assert not (s = '1' and r = '1')
        report "Incorrect use of S_R_flip_flop: s and r both '1'";
    wait on s, r;
end process check;
```

Entities and Passive Processes

We complete this section on behavioral modeling by returning to declarations of entities. We can include certain kinds of concurrent statements in an entity declaration, to monitor use and operation of the entity. The extended syntax rule for an entity declaration that shows this is

entity_declaration ⇐
   entity identifier **is**
      ⟦ **port** (*port*_interface_list) ; ⟧
      { entity_declarative_item }
   ⟦ **begin**
      { concurrent_assertion_statement
      ∥ *passive*_concurrent_procedure_call_statement
      ∥ *passive*_process_statement } ⟧
   end ⟦ **entity** ⟧ ⟦ identifier ⟧ ;

The concurrent statements included in an entity declaration must be *passive*, that is, they may not affect the operation of the entity in any way. A concurrent assertion statement meets this requirement, since it simply tests a condition whenever events occur on signals to which it is sensitive. A process statement is passive if it contains no signal assignment statements or calls to procedures containing signal assignment statements. Such a process can be used to trace events that occur on the entity's inputs. We will describe the remaining alternative, concurrent procedure call statements, when we discuss procedures in Chapter 6. A concurrent procedure call is passive if the procedure called contains no signal assignment statements or calls to procedures containing signal assignment statements.

EXAMPLE

We can rewrite the entity declaration for the set/reset flipflop of Figure 5-22 as shown in Figure 5-23. If we do this, the check is included for every possible implementation of the flipflop and does not need to be included in the corresponding architecture bodies.

FIGURE 5-23

```
entity S_R_flipflop is
    port ( s, r : in bit;  q, q_n : out bit );
begin
    check : assert not (s = '1' and r = '1')
                report "Incorrect use of S_R_flip_flop: s and r both '1'";
end entity S_R_flipflop;
```

The revised entity declaration for the set/reset flipflop, including the concurrent assertion statement to check for correct usage.

EXAMPLE

Figure 5-24 shows an entity declaration for a read-only memory (ROM). It includes a passive process, **trace_reads**, that is sensitive to changes on the **enable** port. When the value of the port changes to '1', the process reports a message tracing the time and address of the read operation. The process does not affect

FIGURE 5-24

```
entity ROM is
    port ( address : in natural;
            data : out bit_vector(0 to 7);
            enable : in bit );
begin
    trace_reads : process (enable) is
    begin
        if enable = '1' then
            report "ROM read at time " & time'image(now)
                        & " from address " & natural'image(address);
        end if;
    end process trace_reads;
end entity ROM;
```

An entity declaration for a ROM, including a passive process for tracing read operations.

the course of the simulation in any way, since it does not include any signal assignments.

5.4 Structural Descriptions

A structural description of a system is expressed in terms of subsystems interconnected by signals. Each subsystem may in turn be composed of an interconnection of sub-subsystems, and so on, until we finally reach a level consisting of primitive components, described purely in terms of their behavior. Thus the top-level system can be thought of as having a hierarchical structure. In this section, we look at how to write structural architecture bodies to express this hierarchical organization.

Component Instantiation and Port Maps

We have seen earlier in this chapter that the concurrent statements in an architecture body describe an implementation of an entity interface. In order to write a structural implementation, we must use a concurrent statement called a *component instantiation* statement, the simplest form of which is governed by the syntax rule

component_instantiation_statement ⇐
    *instantiation*_label :
        entity *entity*_name ⟦ (*architecture*_identifier) ⟧
            ⟦ **port map** (port_association_list) ⟧ ;

This form of component instantiation statement performs *direct instantiation* of an entity. We can think of component instantiation as creating a copy of the named entity, with the corresponding architecture body substituted for the component instance. The port map specifies which ports of the entity are connected to which signals in the enclosing architecture body. The simplified syntax rule for a port association list is

port_association_list $\Leftarrow$
  ([*port*_name =>] (*signal*_name ‖ expression ‖ **open**)) { , ... }

Each element in the association list associates one port of the entity either with one signal of the enclosing architecture body or with the value of an expression, or leaves the port unassociated, as indicated by the keyword **open**.

Let us look at some examples to illustrate component instantiation statements and the association of ports with signals. Suppose we have an entity declared as

```
entity DRAM_controller is
    port ( rd, wr, mem : in bit;
            ras, cas, we, ready : out bit );
end entity DRAM_controller;
```

and a corresponding architecture called fpld. We might create an instance of this entity as follows:

```
main_mem_controller : entity work.DRAM_controller(fpld)
    port map ( cpu_rd, cpu_wr, cpu_mem,
                mem_ras, mem_cas, mem_we, cpu_rdy );
```

In this example, the name **work** refers to the current working library in which entities and architecture bodies are stored. We return to the topic of libraries in the next section. The port map of this example lists the signals in the enclosing architecture body to which the ports of the copy of the entity are connected. *Positional association* is used: each signal listed in the port map is connected to the port at the same position in the entity declaration. So the signal cpu_rd is connected to the port rd, the signal cpu_wr is connected to the port wr and so on.

One of the problems with positional association is that it is not immediately clear which signals are being connected to which ports. Someone reading the description must refer to the entity declaration to check the order of the ports in the entity interface. A better way of writing a component instantiation statement is to use *named association*, as shown in the following example:

```
main_mem_controller : entity work.DRAM_controller(fpld)
    port map ( rd => cpu_rd, wr => cpu_wr,
                mem => cpu_mem, ready => cpu_rdy,
                ras => mem_ras, cas => mem_cas, we => mem_we );
```

Here, each port is explicitly named along with the signal to which it is connected. The order in which the connections are listed is immaterial. The advantage of this approach is that it is immediately obvious to the reader how the entity is connected into the structure of the enclosing architecture body.

In the preceding example we have explicitly named the architecture body to be used corresponding to the entity instantiated. However, the syntax rule for component instantiation statements shows this to be optional. If we wish, we can omit the specification of the architecture body, in which case the one to be used may be chosen when the overall model is processed for simulation, synthesis or some other purpose. At that time, if no other choice is specified, the most recently analyzed architecture body is selected. We return to the topic of analyzing models in the next section.

EXAMPLE

In Figure 5-5 we looked at a behavioral model of an edge-triggered flipflop. We can use the flipflop as the basis of a four-bit edge-triggered register. Figure 5-25 shows the entity declaration and a structural architecture body.

FIGURE 5-25

```
entity reg4 is
    port ( clk, clr, d0, d1, d2, d3 : in bit;
            q0, q1, q2, q3 : out bit );
end entity reg4;

_ _ _ _ _ _ _ _ _ _ _ _ _ _ _ _ _ _ _ _ _ _ _ _ _ _ _ _ _ _ _ _ _ _ _ _ _ _ _ _ _ _

architecture struct of reg4 is
begin
    bit0 : entity work.edge_triggered_Dff(behavioral)
        port map (d0, clk, clr, q0);
    bit1 : entity work.edge_triggered_Dff(behavioral)
        port map (d1, clk, clr, q1);
    bit2 : entity work.edge_triggered_Dff(behavioral)
        port map (d2, clk, clr, q2);
    bit3 : entity work.edge_triggered_Dff(behavioral)
        port map (d3, clk, clr, q3);
end architecture struct;
```

An entity and structural architecture body for a four-bit edge-triggered register, with an asynchronous clear input.

We can use the register entity, along with other entities, as part of a structural architecture for the two-digit decimal counter represented by the schematic of Figure 5-26. Suppose a digit is represented as a bit vector of length four, described by the subtype declaration

```
subtype digit is bit_vector(3 downto 0);
```

Figure 5-27 shows the entity declaration for the counter, along with an outline of the structural architecture body. This example illustrates a number of important points about component instances and port maps. First, the two component instances val0_reg and val1_reg are both instances of the same entity/architecture pair. This means that two distinct copies of the architecture **struct** of **reg4** are created, one for each of the component instances. We return to this point when we discuss the topic of elaboration in the next section. Second, in each of the port maps, ports of the entity being instantiated are associated with separate elements of array signals. This is allowed, since a signal that is of a composite type, such as an array, can be treated as a collection of signals, one per element. Third, some of the signals connected to the component instances are signals declared within the enclosing architecture body, **registered**, whereas the **clk** signal is a port of the entity **reg4**. This again illustrates the point that within an architecture body, the ports of the corresponding entity are treated as signals.

FIGURE 5-26

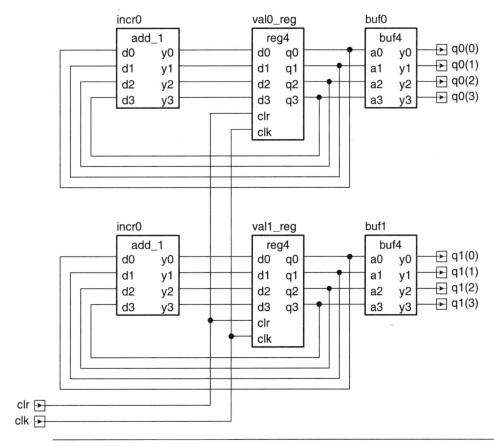

A schematic for a two-digit counter using the reg4 *entity.*

FIGURE 5-27

```
entity counter is
    port ( clk, clr : in bit;  q0, q1 : out digit );
end entity counter;
```

```
architecture registered of counter is
    signal current_val0, current_val1, next_val0, next_val1 : digit;
begin
    val0_reg : entity work.reg4(struct)
        port map ( d0 => next_val0(0), d1 => next_val0(1),
                   d2 => next_val0(2), d3 => next_val0(3),
                   q0 => current_val0(0), q1 => current_val0(1),
                   q2 => current_val0(2), q3 => current_val0(3),
                   clk => clk, clr => clr );
```

(continued on page 140)

(continued from page 139)

```
            val1_reg : entity work.reg4(struct)
                port map ( d0 => next_val1(0), d1 => next_val1(1),
                           d2 => next_val1(2), d3 => next_val1(3),
                           q0 => current_val1(0), q1 => current_val1(1),
                           q2 => current_val1(2), q3 => current_val1(3),
                           clk => clk, clr => clr );
            incr0 : entity work.add_1(boolean_eqn) . . .;
            incr1 : entity work.add_1(boolean_eqn) . . .;
            buf0 : entity work.buf4(basic) . . .;
            buf1 : entity work.buf4(basic) . . .;
        end architecture registered;
```

An entity declaration of a two-digit decimal counter, with an outline of an architecture body using the reg4 *entity.*

We saw in the above example that we can associate separate ports of an instance with individual elements of an actual signal of a composite type, such as an array or record type. If an instance has a composite port, we can write associations the other way around, that is, we can associate separate actual signals with individual elements of the port. This is sometimes called *subelement association*. For example, if the instance DMA_buffer has a port status of type FIFO_status, declared as

```
type FIFO_status is record
        nearly_full, nearly_empty, full, empty : bit;
    end record FIFO_status;
```

we could associate a signal with each element of the port as follows:

```
DMA_buffer : entity work.FIFO
    port map ( . . ., status.nearly_full => start_flush,
                      status.nearly_empty => end_flush,
                      status.full => DMA_buffer_full,
                      status.empty => DMA_buffer_empty, . . . );
```

This illustrates two important points about subelement association. First, all elements of the composite port must be associated with an actual signal. We cannot associate some elements and leave the rest unassociated. Second, all of the associations for a particular port must be grouped together in the association list, without any associations for other ports among them.

We can use subelement association for ports of an array type by writing an indexed element name on the left side of an association. Furthermore, we can associate a slice of the port with an actual signal that is a one-dimensional array, as the following example shows.

EXAMPLE

Suppose we have a register entity, declared as shown at the top of Figure 5-28. The ports **d** and **q** are arrays of bits. The architecture body for a microprocessor, outlined at the bottom of Figure 5-28, instantiates this entity as the program status register (PSR). Individual bits within the register represent condition and interrupt flags, and the field from bit 6 down to bit 4 represents the current interrupt priority level.

FIGURE 5-28

```
entity reg is
    port ( d : in bit_vector(7 downto 0);
            q : out bit_vector(7 downto 0);
            clk : in bit );
end entity reg;

-------------------------------------------------

architecture RTL of microprocessor is
    signal interrupt_req : bit;
    signal interrupt_level : bit_vector(2 downto 0);
    signal carry_flag, negative_flag, overflow_flag, zero_flag : bit;
    signal program_status : bit_vector(7 downto 0);
    signal clk_PSR : bit;
    . . .
begin
    PSR : entity work.reg
        port map ( d(7) => interrupt_req,
                    d(6 downto 4) => interrupt_level,
                    d(3) => carry_flag,    d(2) => negative_flag,
                    d(1) => overflow_flag,  d(0) => zero_flag,
                    q => program_status,
                    clk => clk_PSR );
    . . .
end architecture RTL;
```

An entity declaration for a register with array type ports, and an outline of an architecture body that instantiates the entity. The port map includes subelement associations with individual elements and a slice of the d port.

In the port map of the instance, subelement association is used for the input port **d** to connect individual elements of the port with separate actual signals of the architecture. A slice of the port is connected to the interrupt_level signal. The output port **q**, on the other hand, is associated in whole with the bit-vector signal program_status.

We may also use subelement association for a port that is of an unconstrained array type. The index bounds of the port are determined by the least and greatest index values used in the association list, and the index range direction is determined by the port type. For example, suppose we declare an and gate entity:

```
entity and_gate is
    port ( i : in bit_vector;  y : out bit );
end entity and_gate;
```

and a number of signals:

```
signal serial_select, write_en, bus_clk, serial_wr : bit;
```

We can instantiate the entity as a three-input and gate:

```
serial_write_gate : entity work.and_gate
    port map ( i(1) => serial_select,
               i(2) => write_en,
               i(3) => bus_clk,
               y => serial_wr );
```

Since the input port i is unconstrained, the index values in the subelement associations determine the index bounds for this instance. The least value is one and the greatest value is three. The port type is **bit_vector**, which has an ascending index range. Thus, the index range for the port in the instance is an ascending range from one to three.

The syntax rule for a port association list shows that a port of a component instance may be associated with an expression instead of a signal. In this case, the value of the expression is used as a constant value for the port throughout the simulation. If real hardware is synthesized from the model, the port of the component instance would be tied to a fixed value determined by the expression. Association with an expression is useful when we have an entity provided as part of a library, but we do not need to use all of the functionality provided by the entity. When associating a port with an expression, the value of the expression must be *globally static*, that is, we must be able to determine the value from constants defined when the model is elaborated. So, for example, the expression must not include references to any signals.

EXAMPLE

Given a four-input multiplexer described by the entity declaration

```
entity mux4 is
    port ( i0, i1, i2, i3, sel0, sel1 : in bit;
           z : out bit );
end entity mux4;
```

we can use it as a two-input multiplexer by instantiating it as follows:

```
a_mux : entity work.mux4
    port map ( sel0 => select_line, i0 => line0, i1 => line1,
               z => result_line,
               sel1 => '0', i2 => '1', i3 => '1' );
```

For this component instance, the high-order select bit is fixed at '0', ensuring that only one of **line0** or **line1** is passed to the output. We have also followed the practice, recommended for many logic families, of tying unused inputs to a fixed value, in this case '1'.

Some entities may be designed to allow inputs to be left open by specifying a default value for a port. When the entity is instantiated, we can specify that a port is to be left open by using the keyword **open** in the port association list, as shown in the syntax rule on page 137.

EXAMPLE

The and_or_inv entity declaration on page 105 includes a default value of '1' for each of its input ports, as again shown here:

```
entity and_or_inv is
    port ( a1, a2, b1, b2 : in bit := '1';
            y : out bit );
end entity and_or_inv;
```

We can write a component instantiation to perform the function **not** ((A **and** B) **or** C) using this entity as follows:

```
f_cell : entity work.and_or_inv
    port map ( a1 => A, a2 => B, b1 => C, b2 => open, y => F );
```

The port **b2** is left open, so it assumes the default value '1' specified in the entity declaration.

There is some similarity between specifying a default value for an input port and associating an input port with an expression. In both cases the expression must be globally static (that is, we must be able to determine its value when the model is elaborated). The difference is that a default value is only used if the port is left open when the entity is instantiated, whereas association with an expression specifies that the expression value is to be used to drive the port for the entire simulation or life of the component instance. If a port is declared with a default value and then associated with an expression, the expression value is used, overriding the default value.

Output and bidirectional ports may also be left unassociated using the **open** keyword, provided they are not of an unconstrained array type. If a port of mode **out** is left open, any value driven by the entity is ignored. If a port of mode **inout** is left open, the value used internally by the entity (the *effective value*) is the value that it drives on to the port.

A final point to make about unassociated ports is that we can simply omit a port from a port association list to specify that it remain open. So, given an entity declared as follows:

```
entity and3 is
    port ( a, b, c : in bit := '1';
            z, not_z : out bit );
end entity and3;
```

the component instantiation

g1 : **entity** work.and3 **port map** (a => s1, b => s2, not_z => ctrl1);

has the same meaning as

g1 : **entity** work.and3 **port map** (a => s1, b => s2, not_z => ctrl1,
                              c => **open**, z => **open**);

The difference is that the second version makes it clear that the unused ports are deliberately left open, rather than being accidentally overlooked in the design process. This is useful information for someone reading the model.

VHDL-87

VHDL-87 does not allow direct instantiation. Instead, we must declare a *component* with a similar interface to the entity, instantiate the component and *bind* each component instance to the entity and an associated architecture body. Component declarations and binding are described in Chapter 10.

VHDL-87 does not allow association of an expression with a port in a port map. However, we can achieve a similar effect by declaring a signal, initializing it to the value of the expression and associating the signal with the port. For example, if we declare two signals

signal tied_0 : bit := '0';
signal tied_1 : bit := '1';

we can rewrite the port map shown on page 142 as

port map (sel0 => select_line, i0 => line0, i1 => line1,
              z => result_line,
              sel1 => tied_0, i2 => tied_1, i3 => tied_1);

5.5   Design Processing

Now that we have seen how a design may be described in terms of entities, architectures, component instantiations, signals and processes, it is time to take a practical view. A VHDL description of a design is usually used to simulate the design and perhaps to synthesize the hardware. This involves processing the description using computer-based tools to create a simulation program to run or a hardware net-list to build. Both simulation and synthesis require two preparatory steps: analysis and elaboration. Simulation then involves executing the elaborated model, whereas synthesis involves creating a net-list of primitive circuit elements that perform the same function as the elaborated model. In this section, we look at the analysis, elaboration and execution operations introduced in Chapter 1.

Analysis

The first step in processing a design is to analyze the VHDL descriptions. A correct description must conform to the rules of syntax and semantics that we have discussed at length. An *analyzer* is a tool that verifies this. If a description fails to meet a rule,

the analyzer provides a message indicating the location of the problem and which rule
was broken. We can then correct the error and retry the analysis. Another task performed by the analyzer in most VHDL systems is to translate the description into an
internal form more easily processed by the remaining tools. Whether such a translation
is done or not, the analyzer places each successfully analyzed description into a *design
library*.

A complete VHDL description usually consists of a number of entity declarations
and their corresponding architecture bodies. Each of these is called a *design unit*.
Organizing a design as a hierarchy of modules, rather than as one large flat design, is
good engineering practice. It makes the description much easier to understand and
manage.

The analyzer analyzes each design unit separately and places the internal form into
the library as a *library unit*. If a unit being analyzed uses another unit, the analyzer
extracts information about the other unit from the library, to check that the unit is used
correctly. For example, if an architecture body instantiates an entity, the analyzer
needs to check the number, type and mode of ports of the entity to make sure it is
instantiated correctly. To do this, it requires that the entity be previously analyzed and
stored in the library. Thus, we see that there are dependency relations between library
units in a complete description that enforce an order of analysis of the original design
units.

To clarify this point, we divide design units into *primary units*, which include entity declarations, and *secondary units*, which include architecture bodies. There are other kinds of design units in each class, which we come to in later chapters. A primary
unit defines the external view or interface to a module, whereas a secondary unit describes an implementation of the module. Thus the secondary unit depends on the
corresponding primary unit and must be analyzed after the primary unit has been analyzed. In addition, a library unit may draw upon the facilities defined in some other
primary unit, as in the case of an architecture body instantiating some other entity.
In this case, there is a further dependency between the secondary unit and the refer-
enced primary unit. Thus we may build up a network of dependencies of units upon
primary units. Analysis must be done in such an order that a unit is analyzed before
any of its dependents. Furthermore, whenever we change and reanalyze a primary
unit, all of the dependent units must also be reanalyzed. Note, however, that there
is no way in which any unit can be dependent upon a secondary unit; that is what
makes a secondary unit secondary. This may seem rather complicated, and indeed,
in a large design, the dependency relations can form a complex network. For this reason, most VHDL systems include tools to manage the dependencies, automatically re-
analyzing units where necessary to ensure that an outdated unit is never used.

EXAMPLE

The structural architecture of the **counter** module, described in Figure 5-27,
leads to the network of dependencies shown in Figure 5-29. One possible order
of compilation for this set of design units is

entity **edge_triggered_Dff**
architecture **behav** of **edge_triggered_Dff**

FIGURE 5-29

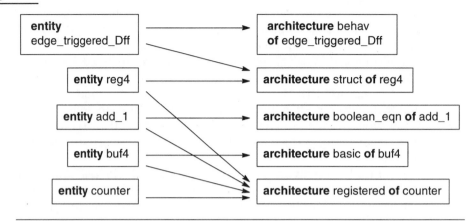

The dependency network for the **counter** *module. The arrows point from a primary unit to a dependent secondary unit.*

entity reg4
architecture struct of reg4

entity add_1
architecture boolean_eqn of add_1

entity buf4
architecture basic of buf

entity counter
architecture registered of counter

In this order, each primary unit is analyzed immediately before its corresponding secondary unit, and each primary unit is analyzed before any secondary unit that instantiates it. This is not the only possible order. Another alternative is to analyze all of the entity declarations first, then analyze the architecture bodies in arbitrary order.

Design Libraries, Library Clauses and Use Clauses

So far, we have not actually said what a design library is, other than that it is where library units are stored. Indeed, this is all that is defined by the VHDL language specification, since to go further is to enter into the domain of the host operating system under which the VHDL tools are run. Some systems may use a database to store analyzed units, whereas others may simply use a directory in the host file system as the design library. The documentation for each VHDL tool suite indicates what we need to know about how the suite deals with design libraries.

A VHDL tool suite must also provide some means of using a number of separate design libraries. When a design is analyzed, we nominate one of the libraries as the *working library,* and the analyzed design is stored in this library. We use the special library name **work** in our VHDL models to refer to the current working library. We have seen examples of this in this chapter's component instantiation statements, in which a previously analyzed entity is instantiated in an architecture body.

If we need to access library units stored in other libraries, we refer to the libraries as *resource libraries*. We do this by including a *library clause* immediately preceding a design unit that accesses the resource libraries. The syntax rule for a library clause is

library_clause ⇐ **library** identifier { , ⚬⚬⚬ } ;

The identifiers are used by the analyzer and the host operating system to locate the design libraries, so that the units contained in them can be used in the description being analyzed. The exact way that the identifiers are used varies between different tool suites and is not defined by the VHDL language specification. Note that we do not need to include the library name **work** in a library clause; the current working library is automatically available.

EXAMPLE

Suppose we are working on part of a large design project code-named Wasp, and we are using standard cell parts supplied by Widget Designs, Inc. Our system administrator has loaded the design library for the Widget cells in a directory called /local/widget/cells in our workstation file system, and our project leader has set up another design library in /projects/wasp/lib for some in-house cells we need to use. We consult the manual for our VHDL analyzer and use operating system commands to set up the appropriate mapping from the identifiers widget_cells and wasp_lib to these library directories. We can then instantiate entities from these libraries, along with entities we have previously analyzed, into our own working library, as shown in Figure 5-30.

FIGURE 5-30

```
library widget_cells, wasp_lib;
architecture cell_based of filter is
    – – declaration of signals, etc
    . . .
begin
    clk_pad : entity wasp_lib.in_pad
        port map ( i => clk, z => filter_clk );
    accum : entity widget_cells.reg32
        port map ( en => accum_en, clk => filter_clk, d => sum,
                    q => result );
    alu : entity work.adder
        port map ( a => alu_op1, b => alu_op2, y => sum, c => carry );
    – – other component instantiations
    . . .
end architecture cell_based;
```

An outline of a library unit referring to entities from the resource libraries widget_cells *and* wasp_lib.

If we need to make frequent reference to library units from a design library, we can include a *use clause* in our model to avoid having to write the library name each time. The simplified syntax rules are

use_clause ⇐ **use** selected_name { , ... } ;

selected_name ⇐ name . (identifier ‖ **all**)

If we include a use clause with a library name as the prefix of the selected name (preceding the dot), and a library unit name from the library as the suffix (after the dot), the library unit is made *directly visible*. This means that subsequent references in the model to the library unit need not prefix the library unit name with the library name. For example, we might precede the architecture body in the previous example with the following library and use clauses:

library widget_cells, wasp_lib;

use widget_cells.reg32;

This makes reg32 directly visible within the architecture body, so we can omit the library name when referring to it in component instantiations; for example:

accum : **entity** reg32
    port map (en => accum_en, clk => filter_clk, d => sum,
              q => result);

If we include the keyword **all** in a use clause, all of the library units within the named library are made directly visible. For example, if we wanted to make all of the Wasp project library units directly visible, we might precede a library unit with the use clause:

use wasp_lib.**all**;

Care should be taken when using this form of use clause with several libraries at once. If two libraries contain library units with the same name, VHDL avoids ambiguity by making neither of them directly visible. The solution is either to use the full selected name to refer to the particular library unit required, or to include in use clauses only those library units really needed in a model.

Use clauses can also be included to make names from packages directly visible. We will return to this idea when we discuss packages in detail in Chapter 7.

Elaboration

Once all of the units in a design hierarchy have been analyzed, the design hierarchy can be *elaborated*. The effect of elaboration is to "flesh out" the hierarchy, producing a collection of processes interconnected by *nets*. This is done by substituting the contents of an architecture body for every instantiation of its corresponding entity. Each net in the elaborated design consists of a signal and the ports of the substituted architecture bodies to which the signal is connected. (Recall that a port of an entity is treated as a signal within a corresponding architecture body.) Let us outline how elaboration proceeds, illustrating it step by step with an example.

Elaboration is a recursive operation, started at the topmost entity in a design hierarchy. We use the **counter** example from Figure 5-27 as our topmost entity. The first step is to create the ports of the entity. Next, an architecture body corresponding to the entity is chosen. If we do not explicitly specify which architecture body to choose, the most recently analyzed architecture body is used. For this illustration, we use the architecture **registered**. This architecture body is then elaborated, first by creating any signals that it declares, then by elaborating each of the concurrent statements in its body. Figure 5-31 shows the **counter** design with the signals created.

FIGURE 5-31

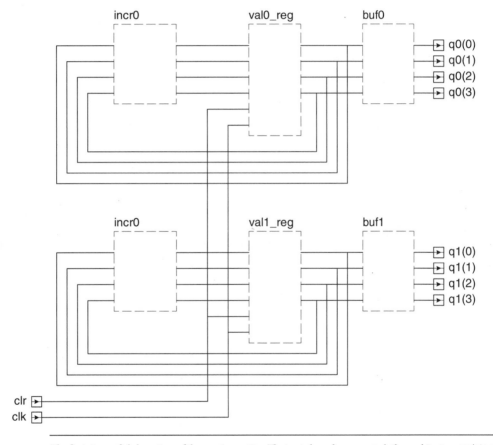

The first stage of elaboration of the **counter** *entity. The ports have been created, the architecture* **registered** *selected and the signals of the architecture created.*

The concurrent statements in this architecture are all component instantiation statements. Each of them is elaborated by creating new instances of the ports specified by the instantiated entity and joining them into the nets represented by the signals with which they are associated. Then the internal structure of the specified architecture body of the instantiated entity is copied in place of the component instance, as shown in Figure 5-32. The architectures substituted for the instances of the **add_1** and **buf4** entities are both behavioral, consisting of processes that read the input ports and make assignments to the output ports. Hence elaboration is complete for these architectures. However, the architecture **struct**, substituted for each of the instances of **reg4**, contains

FIGURE 5-32

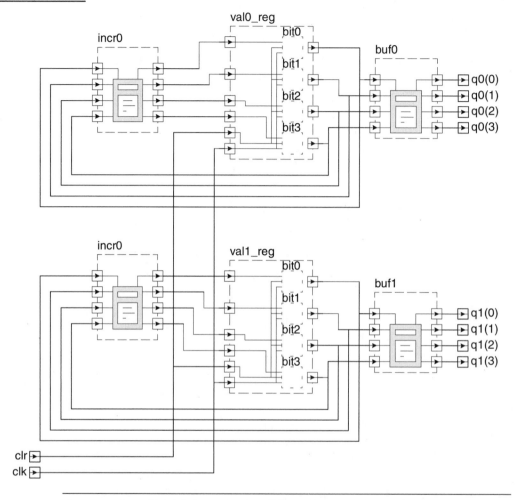

The counter design further elaborated. Behavioral architectures, consisting of just processes, have been substituted for instances of the add_1 *and* buf4 *entities. A structural architecture has been substituted for each instance of the* reg4 *entity.*

further signals and component instances. Hence they are elaborated in turn, producing the structure shown in Figure 5-33 for each instance. We have now reached a stage where we have a collection of nets comprising signals and ports, and processes that sense and drive the nets.

Each process statement in the design is elaborated by creating new instances of the variables it declares and by creating a driver for each of the signals for which it has signal assignment statements. The drivers are joined to the nets containing the signals they drive. For example, the **storage** process within **bit0** of **val0_reg** has a driver for the port **q**, which is part of the net based on the signal **current_val0(0)**.

Once all of the component instances and all of the resulting processes have been elaborated, elaboration of the design hierarchy is complete. We now have a fully

FIGURE 5-33

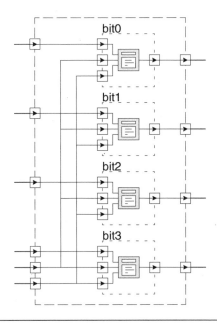

A register within the counter structure elaborated down to architectures that consist only of processes and signals.

fleshed-out version of the design, consisting of a number of process instances and a number of nets connecting them. Note that there are several distinct instances of some of the processes, one for each use of an entity containing the process, and each process instance has its own distinct version of the process variables. Each net in the elaborated design consists of a signal, a collection of ports associated with it and a driver within a process instance.

Execution

Now that we have an elaborated design hierarchy, we can execute it to simulate operation of the system it describes. Much of our previous discussion of VHDL statements was in terms of what happens when they are executed, so we do not go over statement execution again here. Instead, we concentrate on the simulation algorithm introduced in Chapter 1.

Recall that the simulation algorithm consists of an initialization phase followed by a repeated simulation cycle. The simulator keeps a clock to measure out the passage of simulation time. In the initialization phase, the simulation time is set to zero. Each driver is initialized to drive its signal with the initial value declared for the signal or the default value for the signal if no initial value was declared. Next, each of the process instances in the design is started and executes the sequential statements in its body. We usually write a model so that at least some of these initial statements schedule some transactions to get the simulation under way, then suspend by executing a wait statement. When all of the process instances have suspended, initialization is complete and the simulator can start the first simulation cycle.

At the beginning of a simulation cycle, there may be a number of drivers with transactions scheduled on them and a number of process instances that have scheduled timeouts. The first step in the simulation cycle is to advance the simulation time clock to the earliest time at which a transaction or process timeout has been scheduled. Second, all of the transactions scheduled for this time are performed, updating the corresponding signals and possibly causing events on those signals. Third, all process instances that are sensitive to any of these events are resumed. In addition, process instances whose timeout expires at the current simulation time are resumed during this step. All of these processes execute their sequential statements, possibly scheduling more transactions or timeouts, and eventually suspend again by executing wait statements. When they have all suspended, the simulation cycle is done and the next cycle can start. If there are no more transactions or timeouts scheduled, or if simulation time reaches **time'high** (the largest representable time value), the simulation is complete.

Describing the operation of a simulator in this way is a little like setting a play in a theatre without any seats—nobody is there to watch it, so what's the point! In reality, a simulator is part of a suite of VHDL tools and provides us with various means to control and monitor the progress of the simulation. Typical simulators allow us to step through the model one line at a time or to set breakpoints, causing the simulation to stop when a line of the model is executed or a signal is assigned a particular value. They usually provide commands to display the value of signals or variables. Many simulators also provide a graphical waveform display of the history of signal values similar to a logic analyzer display, and allow storage and subsequent redisplay of the history for later analysis. It is these facilities that make the simulation useful. Unfortunately, since there is a great deal of variation between the facilities provided by different simulators, it is not practical to go into any detail in this book. Simulator vendors usually provide training documentation and lab courses that explain how to use the facilities provided by their products.

Exercises

1. [❶ 5.1] Write an entity declaration for a lookup table ROM modeled at an abstract level. The ROM has an address input of type **lookup_index**, which is an integer range from 0 to 31, and a data output of type **real**. Include declarations within the declarative part of the entity to define the ROM contents, initialized to numbers of your choice.

2. [❶ 5.3] Trace the transactions applied to the signal **s** in the following process. At what times is the signal active, and at what times does an event occur on it?

```
process is
begin
    s <= 'Z', '0' after 10 ns, '1' after 30 ns;
    wait for 50 ns;
    s <= '1' after 5 ns; 'H' after 15 ns;
    wait for 50 ns;
    s <= 'Z';
    wait;
end process;
```

3. [❶ 5.3] Given the assignments to the signal s made by the process in Exercise 2, trace the values of the signals s'delayed(5 ns), s'stable(5 ns), s'quiet(5 ns) and s'transaction. What are the values of s'last_event, s'last_active and s'last_value at time 60 ns?

4. [❶ 5.3] Write a wait statement that suspends a process until a signal s changes from '1' to '0' while an enable signal en is '1'.

5. [❶ 5.3] Write a wait statement that suspends a process until a signal **ready** changes to '1' or until a maximum of 5 ms has elapsed.

6. [❶ 5.3] Suppose the signal s currently has the value '0'. What is the value of the Boolean variables v1 and v2 after execution of the following statements within a process?

```
s <= '1';
v1 := s = '1';
wait on s;
v2 := s = '1';
```

7. [❶ 5.3] Trace the transactions scheduled on the driver for z by the following statements, and show the values taken on by z during simulation.

```
z <= transport '1' after 6 ns;
wait for 3 ns;
z <= transport '0' after 4 ns;
wait for 5 ns;
z <= transport '1' after 6 ns;
wait for 1 ns;
z <= transport '0' after 4 ns;
```

8. [❶ 5.3] Trace the transactions scheduled on the driver for x by the following statements, and show the values taken on by x during simulation. Assume x initially has the value zero.

```
x <= reject 5 ns inertial 1 after 7 ns, 23 after 9 ns, 5 after 10 ns,
                  23 after 12 ns, -5 after 15 ns;
wait for 6 ns;
x <= reject 5 ns inertial 23 after 7 ns;
```

9. [❶ 5.3] Write the equivalent process for the conditional signal assignment statement

```
mux_logic :
    z <= a and not b after 5 ns when enable = '1' and sel = '0' else
         x or y after 6 ns when enable = '1' and sel = '1' else
         '0' after 4 ns;
```

10. [❶ 5.3] Write the equivalent process for the selected signal assignment statement

```
with bit_vector'(s, r) select
    q <= unaffected when "00",
         '0' when "01",
         '1' when "10" | "11";
```

11. [❶ 5.3] Write a concurrent assertion statement that verifies that the time between changes of a clock signal, clk, is at least T_pw_clk.

12. [❶ 5.4] Write component instantiation statements to model the structure shown by the following schematic diagram. Assume that the entity **ttl_74x74** and the corresponding architecture **basic** have been analyzed into the library **work**.

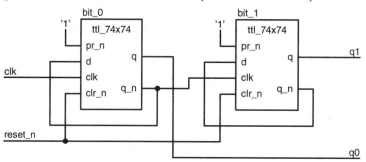

13. [❶ 5.4] Sketch a schematic diagram of the structure modeled by the following component instantiation statements.

```
decode_1 : entity work.ttl_74x138(basic)
    port map ( c => a(2), b => a(1), a => a(0),
               g1 => a(3), g2a_n => sel_n, g2b_n => '0',
               y7_n => en_n(15), y6_n => en_n(14),
               y5_n => en_n(13), y4_n => en_n(12),
               y3_n => en_n(11), y2_n => en_n(10),
               y1_n => en_n(9), y0_n => en_n(8) );

decode_0 : entity work.ttl_74x138(basic)
    port map ( c => a(2), b => a(1), a => a(0),
               g1 => '1', g2a_n => sel_n, g2b_n => a(3),
               y7_n => en_n(7), y6_n => en_n(6),
               y5_n => en_n(5), y4_n => en_n(4),
               y3_n => en_n(3), y2_n => en_n(2),
               y1_n => en_n(1), y0_n => en_n(0) );
```

14. [❶ 5.5] The example on page 145 shows one possible order of analysis of the design units in the counter of Figure 5-29. Show two other possible orders of analysis.

15. [❶ 5.5] Write a context clause that makes the resource libraries **company_lib** and **project_lib** accessible and that makes directly visible the entities **in_pad** and **out_pad** from **company_lib** and all entities from **project_lib**.

16. [❷ 5.3] Develop a behavioral model for a four-input multiplexer, with ports of type bit and a propagation delay from data or select input to data output of 4.5 ns. You should declare a constant for the propagation delay, rather than writing it as a literal in signal assignments in the model.

17. [❷ 5.3] Develop a behavioral model for a negative-edge-triggered four-bit counter with asynchronous parallel load inputs. The entity declaration is

```
entity counter is
    port ( clk_n, load_en : in std_ulogic;
           d : in std_ulogic_vector(3 downto 0);
           q : out std_ulogic_vector(3 downto 0) );
end entity counter;
```

18. [❷ 5.3] Develop a behavioral model for a D-latch with a clock-to-output propagation delay of 3 ns and a data-to-output propagation delay of 4 ns.

19. [❷ 5.3] Develop a behavioral model for an edge-triggered flipflop that includes tests to verify the following timing constraints: data setup time of 3 ns, data hold time of 2 ns and minimum clock pulse width of 5 ns.

20. [❷ 5.3] Develop a model of an adder whose interface is specified by the following entity declaration:

```
entity adder is
    port ( a, b : in integer;  s : out integer );
end entity adder;
```

For each pair of integers that arrive on the inputs, the adder produces their sum on the output. Note that successive integers on each input may have the same value, so the adder must respond to transactions rather than to events. While integers in a pair may arrive in the inputs at different times, you may assume that neither value of the following pair will arrive until both values of the first pair have arrived. The adder should produce the sum only when both input values of a pair have arrived.

21. [❷ 5.3] Develop a behavioral model for a two-input Muller-C element, with two input ports and one output, all of type **bit**. The inputs and outputs are initially '0'. When both inputs are '1', the output changes to '1'. It stays '1' until both inputs are '0', at which time it changes back to '0'. Your model should have a propagation delay for rising output transitions of 3.5 ns, and for falling output transitions of 2.5 ns.

22. [❷ 5.3] The following process statement models a producer of data:

```
producer : process is
    variable next_data : natural := 0;
begin
    data <= next_data; next_data := next_data + 1;
    data_ready <= '1';
    wait until data_ack = '1';
    data read <= '0';
    wait until data_ack = '0';
end process producer;
```

The process uses a four-phase handshaking protocol to synchronize data transfer with a consumer process. Develop a process statement to model the consumer. It, too, should use delta delays in the handshaking protocol. Include the process statements in a test-bench architecture body, and experiment with your simulator to see how it deals with models that use delta delays.

23. [❷ 5.3] Develop a behavioral model for a multitap delay line, with the following interface:

```
entity delay_line is
    port ( input : in std_ulogic;  output : out std_ulogic_vector );
end entity delay_line;
```

Each element of the output port is a delayed version of the input. The delay to the leftmost output element is 5 ns, to the next element is 10 ns and so on. The delay to the rightmost element is 5 ns times the length of the output port. Assume the delay line acts as an ideal transmission line.

24. [❷ 5.3] Develop a functional model using conditional signal assignment statements of an address decoder for a microcomputer system. The decoder has an address input port of type **natural** and a number of active-low select outputs, each activated when the address is within a given range. The outputs and their corresponding ranges are

> ROM_sel_n   16#0000# to 16#3FFF#
> RAM_sel_n   16#4000# to 16#5FFF#
> PIO_sel_n   16#8000# to 16#8FFF#
> SIO_sel_n   16#9000# to 16#9FFF#
> INT_sel_n   16#F000# to 16#FFFF#

25. [❷ 5.3] Develop a functional model of a BCD-to-seven-segment decoder for a light-emitting diode (LED) display. The decoder has a four-bit input that encodes a numeric digit between 0 and 9. There are seven outputs indexed from 'a' to 'g', corresponding to the seven segments of the LED display as shown in the margin. An output bit being '1' causes the corresponding segment to illuminate. For each input digit, the decoder activates the appropriate combination of segment outputs to form the displayed representation of the digit. For example, for the input "0010", which encodes the digit 2, the output is "1101101". Your model should use a selected signal assignment statement to describe the decoder function in truth-table form.

26. [❷ 5.3] Write an entity declaration for a four-bit counter with an asynchronous reset input. Include a process in the entity declaration that measures the duration of each reset pulse and reports the duration at the end of each pulse.

27. [❷ 5.4] Develop a structural model of an eight-bit odd-parity checker using instances of an exclusive-or gate entity. The parity checker has eight inputs, i0 to i7, and an output, p, all of type **std_ulogic**. The logic equation describing the parity checker is

$$P = ((I_0 \oplus I_1) \oplus (I_2 \oplus I_3)) \oplus ((I_4 \oplus I_5) \oplus (I_6 \oplus I_7))$$

28. [❸ 5.4] Develop a structural model of a 14-bit counter with parallel load inputs, using instances of the four-bit counter described in Exercise 17. Ensure that any unused inputs are properly connected to a constant driving value.

29. [❸ 5.3] Develop a behavioral model for a D-latch with tristate output. The entity declaration is

> **entity** d_latch **is**
>   **port** (latch_en, out_en, d : **in** std_ulogic; q : **out** std_ulogic);
> **end entity** d_latch;

When latch_en is asserted, data from the d input enters the latch. When latch_en is negated, the latch maintains the stored value. When out_en is asserted, data passes through to the output. When out_en is negated, the output has the value 'Z' (high-impedance). The propagation delay from latch_en to q is 3 ns and from d to q is 4 ns. The delay from out_en asserted to q active is 2 ns and from out_en negated to q high-impedance is 5 ns.

30. [❸ 5.3] Develop a functional model of a four-bit carry-look-ahead adder. The adder has two four-bit data inputs, a(3 **downto** 0) and b(3 **downto** 0), a four-bit data output, s(3 **downto** 0), a carry input, c_in, a carry output, c_out, a carry generate output, g and a carry propagate output, p. The adder is described by the logic equations and associated propagation delays:

$$S_i = A_i \oplus B_i \oplus C_{i-1} \text{ (delay is 5 ns)}$$
$$G_i = A_i B_i \text{ (delay is 2 ns)}$$
$$P_i = A_i + B_i \text{ (delay is 3 ns)}$$
$$C_i = G_i + P_i C_{i-1} = G_i + P_i G_{i-1} + P_i P_{i-1} G_{i-2} + \ldots + P_i P_{i-1} \ldots P_0 C_{-1}$$
$$\text{(delay is 5 ns)}$$
$$G = G_3 + P_3 G_2 + P_3 P_2 G_1 + P_3 P_2 P_1 G_0 \text{ (delay is 5 ns)}$$
$$P = P_3 P_2 P_1 P_0 \text{ (delay is 3 ns)}$$

where the G_i are the intermediate carry generate signals, the P_i are the intermediate carry propagate signals and the C_i are the intermediate carry signals. C_{-1} is c_in and C_3 is c_out. Your model should use the expanded equation to calculate the intermediate carries, which are then used to calculate the sums.

31. [❸ 5.3] Develop a behavioral model for a four-input arbiter with the following entity interface:

```
entity arbiter is
    port ( request : in bit_vector(0 to 3);
           acknowledge : out bit_vector(0 to 3) );
end entity arbiter;
```

The arbiter should use a round-robin discipline for responding to requests. Augment the entity declaration by including a concurrent assertion statement that verifies that no more than one acknowledgement is issued at once and that an acknowledgement is only issued to a requesting client.

32. [❸ 5.3] Write an entity declaration for a 7474 positive edge-triggered JK-flipflop with asynchronous active-low preset and clear inputs, and Q and $\overline{Q}$ outputs. Include concurrent assertion statements and passive processes as necessary in the entity declaration to verify that

   • the preset and clear inputs are not activated simultaneously,
   • the setup time of 6 ns from the J and K inputs to the rising clock edge is observed,
   • the hold time of 2 ns for the J and K inputs after the rising clock edge is observed, and
   • the minimum pulse width of 5 ns on each of the clock, preset and clear inputs is observed.

   Write a behavioral architecture body for the flipflop and a test bench that exercises the statements in the entity declaration.

33. [❸ 5.4] Define entity interfaces for a microprocessor, a ROM, a RAM, a parallel I/O controller, a serial I/O controller, an interrupt controller and a clock generator. Use instances of these entities and an instance of the address decoder described in Exercise 24 to develop a structural model of a microcomputer system.

34. [❸ 5.4] Develop a structural model of a 16-bit carry-look-ahead adder, using instances of the four-bit adder described in Exercise 30. You will need to develop a carry-look-ahead generator with the following interface:

> **entity** carry_look_ahead_generator **is**
> **port** (p0, p1, p2, p3, g0, g1, g2, g3 : **in** bit;
>   c_in : **in** bit; c1, c2, c3 : **out** bit);
> **end entity** carry_look_ahead_generator

The carry-look-ahead generator is connected to the four-bit adders as follows:

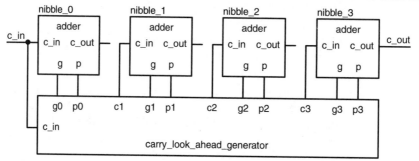

It calculates the carry output signals using the generate, propagate and carry inputs in the same way that the four-bit counters calculate their internal carry signals.

Subprograms

When we write complex behavioral models it is useful to divide the code into sections, each dealing with a relatively self-contained part of the behavior. VHDL provides a *subprogram* facility to let us do this. In this chapter, we look at the two kinds of subprograms: *procedures* and *functions*. The difference between the two is that a procedure encapsulates a collection of sequential statements that are executed for their effect, whereas a function encapsulates a collection of statements that compute a result. Thus a procedure is a generalization of a statement, whereas a function is a generalization of an expression.

6.1 Procedures

We start our discussion of subprograms with procedures. There are two aspects to using procedures in a model: first the procedure is declared, then elsewhere the procedure is *called*. The syntax rule for a procedure declaration is

> subprogram_body ⇐
>   **procedure** identifier ⟦ (*parameter*_interface_list) ⟧ **is**
>   { subprogram_declarative_part }
>   **begin**
>   { sequential_statement }
>   **end** ⟦ **procedure** ⟧ ⟦ identifier ⟧ ;

For now we will just look at procedures without the parameter list part; we will come back to parameters in the next section.

The *identifier* in a procedure declaration names the procedure. The name may be repeated at the end of the procedure declaration. The sequential statements in the body of a procedure implement the algorithm that the procedure is to perform and can include any of the sequential statements that we have seen in previous chapters. A procedure can declare items in its declarative part for use in the statements in the procedure body. The declarations can include types, subtypes, constants, variables and nested subprogram declarations. The items declared are not accessible outside of the procedure; we say they are *local* to the procedure.

EXAMPLE

Figure 6-1 is a declaration for a procedure that calculates an average of a collection of data values stored in an array called **samples** and assigns the result to a variable called **average**. This procedure has a local variable **total** for accumulating the sum of array elements. Unlike variables in processes, procedure local variables are created anew and initialized each time the procedure is called.

FIGURE 6-1

```
procedure average_samples is
    variable total : real := 0.0;
begin
    assert samples'length > 0 severity failure;
    for index in samples'range loop
        total := total + samples(index);
    end loop;
    average := total / real(samples'length);
end procedure average_samples;
```

A declaration for a procedure to average a number of values.

The actions of a procedure are invoked by a *procedure call* statement, which is yet another VHDL sequential statement. A procedure with no parameters is called simply by writing its name, as shown by the syntax rule

procedure_call_statement ⇐ ⟦ label : ⟧ *procedure*_name ;

The optional label allows us to identify the procedure call statement. As an example, we might include the following statement in a process:

average_samples;

The effect of this statement is to invoke the procedure **average_samples**. This involves creating and initializing a new instance of the local variable **total**, then executing the statements in the body of the procedure. When the last statement in the procedure is completed, we say the procedure *returns*; that is, the thread of control of statement execution returns to the process from which the procedure was called, and the next statement in the process after the call is executed.

We can write a procedure declaration in the declarative part of an architecture body or a process. We can also declare procedures within other procedures, but we will leave that until a later section. If a procedure is included in an architecture body's declarative part, it can be called from within any of the processes in the architecture body. On the other hand, declaring a procedure within a process hides it away from use by other processes.

EXAMPLE

The outline in Figure 6-2 illustrates a procedure defined within a process. The procedure **do_arith_op** encapsulates an algorithm for arithmetic operations on two values, producing a result and a flag indicating whether the result is zero. It has a variable **result**, which it uses within the sequential statements that implement the algorithm. The statements also use the signals and other objects declared in the architecture body. The process **alu** invokes **do_arith_op** with a procedure call statement. The advantage of separating the statements for arithmetic operations into a procedure in this example is that it simplifies the body of the **alu** process.

FIGURE 6-2

```
architecture rtl of control_processor is
    type func_code is (add, subtract);
    signal op1, op2, dest : integer;
    signal Z_flag : boolean;
    signal func : func_code;
        . . .
begin
```

(continued on page 162)

(continued from page 161)

```
alu : process is
    procedure do_arith_op is
        variable result : integer;
    begin
        case func is
            when add =>
                result := op1 + op2;
            when subtract =>
                result := op1 – op2;
        end case;
        dest  <=  result after Tpd;
        Z_flag  <=  result = 0 after Tpd;
    end procedure do_arith_op;
begin
    . . .
    do_arith_op;
    . . .
end process alu;

. . .

end architecture rtl;
```

An outline of an architecture body with a process containing a procedure. The procedure encapsulates part of the behavior of the process and is invoked by the procedure call statement within the process.

Another important use of procedures arises when some action needs to be performed several times at different places in a model. Instead of writing several copies of the statements to perform the action, the statements can be encapsulated in a procedure, which is then called from each place.

EXAMPLE

Figure 6-3 shows an outline of a process taken from a behavioral model of a CPU. The process fetches instructions from memory and interprets them. Since the actions required to fetch an instruction and to fetch a data word are identical, the process encapsulates them in a procedure, read_memory. The procedure copies the address from the memory address register to the address bus, sets the memory read signal to '1', then activates the memory request signal. When the memory responds, the procedure copies the data from the data bus signal to the memory data register and acknowledges to the memory by setting the request signal back to '0'. When the memory has completed its operation, the procedure returns.

The procedure is called in two places within the process. First, it is called to fetch an instruction. The process copies the program counter into the memory address register and calls the procedure. When the procedure returns, the process copies the data from the memory data register, placed there by the procedure, to the instruction register. The second call to the procedure takes place when a "load

FIGURE 6-3 _____

```
instruction_interpreter : process is
    variable mem_address_reg, mem_data_reg,
            prog_counter, instr_reg, accumulator, index_reg : word;
    . . .
    procedure read_memory is
    begin
        address_bus <= mem_address_reg;
        mem_read <= '1';
        mem_request <= '1';
        wait until mem_ready = '1';
        mem_data_reg := data_bus_in;
        mem_request <= '0';
        wait until mem_ready = '0';
    end procedure read_memory;
begin
    . . .        - - initialization
    loop
        - - fetch next instruction
        mem_address_reg := prog_counter;
        read_memory;                         - - call procedure
        instr_reg := mem_data_reg;
        . . .
        case opcode is
            . . .
            when load_mem =>
                mem_address_reg := index_reg + displacement;
                read_memory;                  - - call procedure
                accumulator := mem_data_reg;
            . . .
        end case;
    end loop;
end process instruction_interpreter;
```

An outline of an instruction interpreter process from a CPU model. The procedure read_memory *is called from two places.*

memory" instruction is executed. The process sets the memory address register using the values of the index register and some displacement, then calls the memory read procedure to perform the read operation. When it returns, the process copies the data to the accumulator.

Since a procedure call is a form of sequential statement and a procedure body implements an algorithm using sequential statements, there is no reason why one procedure cannot call another procedure. In this case, control is passed from the calling procedure to the called procedure to execute its statements. When the called procedure returns, the calling procedure carries on executing statements until it returns to its caller.

EXAMPLE

The process outlined in Figure 6-4 is a control sequencer for a register-transfer-level model of a CPU. It sequences the activation of control signals with a two-phase clock on signals **phase1** and **phase2**. The process contains two procedures, **control_write_back** and **control_arith_op**, that encapsulate parts of the control algorithm. The process calls **control_arith_op** when an arithmetic operation must be performed. This procedure sequences the control signals for the source and destination operand registers in the data path. It then calls **control_write_back**, which sequences the control signals for the register file in the data path, to write the value from the destination register. When this procedure is completed, it returns to the first procedure, which then returns to the process.

FIGURE 6-4

```
control_sequencer : process is
    procedure control_write_back is
    begin
        wait until phase1 = '1';
        reg_file_write_en <= '1';
        wait until phase2 = '0';
        reg_file_write_en <= '0';
    end procedure control_write_back;

    procedure control_arith_op is
    begin
        wait until phase1 = '1';
        A_reg_out_en <= '1';
        B_reg_out_en <= '1';
        wait until phase1 = '0';
        A_reg_out_en <= '0';
        B_reg_out_en <= '0';
        wait until phase2 = '1';
        C_reg_load_en <= '1';
        wait until phase2 = '0';
        C_reg_load_en <= '0';
        control_write_back;            -- call procedure
    end procedure control_arith_op;
    . . .
begin
    . . .
    control_arith_op;                  -- call procedure
    . . .
end process control_sequencer;
```

An outline of a control sequencer processor for a register-transfer-level model of a CPU. The process contains procedures that encapsulate parts of the control algorithm. The process calls these procedures, which may in turn call other procedures.

VHDL-87

The keyword **procedure** may not be included at the end of a procedure declaration in VHDL-87. Procedure call statements may not be labeled in VHDL-87.

Return Statement in a Procedure

In all of the examples above, the procedures completed execution of the statements in their bodies before returning. Sometimes it is useful to be able to return from the middle of a procedure, for example, as a way of handling an exceptional condition. We can do this using a *return* statement, described by the simplified syntax rule

return_statement ⇐ ⟦ label : ⟧ **return** ;

The optional label allows us to identify the return statement. The effect of the return statement, when executed in a procedure, is that the procedure is immediately terminated and control is transferred back to the caller.

EXAMPLE

Figure 6-5 is a revised version of the instruction interpreter process from Figure 6-3. The procedure to read from memory is revised to check for the reset signal becoming active during a read operation. If it does, the procedure returns immediately, aborting the operation in progress. The process then exits the fetch/execute loop and starts the process body again, reinitializing its state and output signals.

FIGURE 6-5

```
instruction_interpreter : process is

    . . .

    procedure read_memory is
    begin
        address_bus <= mem_address_reg;
        mem_read <= '1';
        mem_request <= '1';
        wait until mem_ready = '1' or reset = '1';
        if reset = '1' then
            return;
        end if;
        mem_data_reg := data_bus_in;
        mem_request <= '0';
        wait until mem_ready = '0';
    end procedure read_memory;
```

(continued on page 166)

(continued from page 165)

```
        begin
            . . .         – – initialization
        loop
            . . .
            read_memory;
            exit when reset = '1';
            . . .
        end loop;
    end process instruction_interpreter;
```

A revised instruction interpreter process. The read memory procedure now checks for the reset signal becoming active.

VHDL-87

Return statements may not be labeled in VHDL-87.

6.2 Procedure Parameters

Now that we have looked at the basics of procedures, we will discuss procedures that include parameters. A *parameterized procedure* is much more general in that it can perform its algorithm using different data objects or values each time it is called. The idea is that the caller passes parameters to the procedure as part of the procedure call, and the procedure then executes its statements using the parameters.

When we write a parameterized procedure, we include information in the *parameter interface list* (or *parameter list*, for short) about the parameters to be passed to the procedure. The syntax rule for a procedure declaration on page 160 shows where the parameter list fits in. Following is the syntax rule for a parameter list:

interface_list ⇐
    ([**constant** ‖ **variable** ‖ **signal**]
      identifier { , ₀₀₀ } : [mode] subtype_indication
                           [:= *static*_expression]) { ; ₀₀₀ }
    mode ⇐ **in** ‖ **out** ‖ **inout**

As we can see, it is similar to the port interface list used in declaring entities. This similarity is not coincidental, since they both specify information about objects upon which the user and the implementation must agree. In the case of a procedure, the user is the caller of the procedure, and the implementation is the body of statements within the procedure. The objects defined in the parameter list are called the *formal parameters* of the procedure. We can think of them as placeholders that stand for the *actual parameters*, which are to be supplied by the caller when it calls the procedure. Since the syntax rule for a parameter list is quite complex, let us start with some simple examples and work up from them.

EXAMPLE

First, let's rewrite the procedure **do_arith_op**, from Figure 6-2 on pages 161–162, so that the function code is passed as a parameter. The new version is shown in Figure 6-6. In the parameter interface list we have identified one formal parameter named **op**. This name is used in the statements in the procedure to refer to the value that will be passed as an actual parameter when the procedure is called. The mode of the formal parameter is **in**, indicating that it is used to pass information into the procedure from the caller. This means that the statements in the procedure can use the value but cannot modify it. In the parameter list we have specified the type of the parameter as func_code. This indicates that the operations performed on the value in the statements must be appropriate for a value of this type, and that the caller may only pass a value of this type as an actual parameter.

FIGURE 6-6

```
procedure do_arith_op ( op : in func_code ) is
    variable result : integer;
begin
    case op is
        when add =>
            result := op1 + op2;
        when subtract =>
            result := op1 – op2;
    end case;
    dest  <=  result after Tpd;
    Z_flag  <=  result = 0 after Tpd;
end procedure do_arith_op;
```

A procedure to perform an arithmetic operation, parameterized by the kind of operation.

Now that we have parameterized the procedure, we can call it from different places passing different function codes each time. For example, a call at one place might be

do_arith_op (add);

The procedure call simply includes the actual parameter value in parentheses. In this case we pass the literal value **add** as the actual parameter. At another place in the model we might pass the value of the signal func shown in the model on pages 161–162:

do_arith_op (func);

In this example, we have specified the mode of the formal parameter as **in**. Note that the syntax rule for a parameter list indicates that the mode is an optional part. If we leave it out, mode **in** is assumed, so we could have written the procedure as

procedure do_arith_op (op : func_code) **is** . . .

While this is equally correct, it's not a bad idea to include the mode specification for **in** parameters, to make our intention explicitly clear.

The syntax rule for a parameter list also shows us that we can specify the *class* of a formal parameter, namely, whether it is a constant, a variable or a signal within the procedure. If the mode of the parameter is **in**, the class is assumed to be *constant*, since a constant is an object that cannot be updated by assignment. It is just a quirk of VHDL that we can specify both **constant** and **in**, even though to do so is redundant. Usually we simply leave out the keyword **constant**, relying on the mode to make our intentions clear. For an **in** mode constant-class parameter, we write an expression as the actual parameter. The value of this expression must be of the type specified in the parameter list. The value is passed to the procedure for use in the statements in its body.

Let us now turn to formal parameters of mode **out**. Such a parameter lets us transfer information out from the procedure back to the caller. Here is an example, before we delve into the details.

EXAMPLE

The procedure in Figure 6-7 performs addition of two unsigned numbers represented as bit vectors of type **word32**, which we assume is defined elsewhere. The procedure has two **in** mode parameters **a** and **b**, allowing the caller to pass two bit-vector values. The procedure uses these values to calculate the sum and overflow flag. Within the procedure, the two **out** mode parameters, **result** and **overflow**, appear as variables. The procedure performs variable assignments to update their values, thus transferring information back to the caller.

FIGURE 6-7

```
procedure addu ( a, b : in word32;
                        result : out word32;  overflow : out boolean ) is
    variable sum : word32;
    variable carry : bit := '0';
begin
    for index in sum'reverse_range loop
        sum(index) := a(index) xor b(index) xor carry;
        carry := ( a(index) and b(index) )
                        or ( carry and ( a(index) xor b(index) ) );
    end loop;
    result := sum;
    overflow := carry = '1';
end procedure addu;
```

A procedure to add two bit vectors representing unsigned integers.

A call to this procedure may appear as follows:

```
variable PC, next_PC : word32;
variable overflow_flag : boolean;
    . . .
```

addu (PC, X"0000_0004", next_PC, overflow_flag);

In this procedure call statement, the first two actual parameters are expressions, whose values are passed in through the formal parameters a and b. The third and fourth actual parameters are the names of variables. When the procedure returns, the values assigned by the procedure to the formal parameters result and overflow are used to update the variables next_PC and overflow_flag.

In the above example, the **out** mode parameters are of the class *variable*. Since this class is assumed for **out** parameters, we usually leave out the class specification **variable**, although it may be included if we wish to state the class explicitly. We will come back to signal-class parameters in a moment. The mode **out** indicates that the only way the procedure may use the formal parameters is to update them by variable assignment to transfer information back to the caller. It may not read the parameter values, as it can with **in** mode parameters. For an **out** mode, variable-class parameter, the caller must supply a variable as an actual parameter. Both the actual parameter and the value returned must be of the type specified in the parameter list.

The third mode we can specify for formal parameters is **inout**, which is a combination of **in** and **out** modes. It is used for objects that are to be both read and updated by a procedure. As with **out** parameters, they are assumed to be of class variable if the class is not explicitly stated. For **inout** mode variable parameters, the caller supplies a variable as an actual parameter. The value of this variable is used to initialize the formal parameter, which may then be used in the statements of the procedure. The procedure may also perform variable assignments to update the formal parameter. When the procedure returns, the value of the formal parameter is copied back to the actual parameter variable, transferring information back to the caller.

EXAMPLE

The procedure in Figure 6-8 negates a number represented as a bit vector, using the "complement and add one" method. Since a is an **inout** mode parameter, we can refer to its value in expressions in the procedure body. (This differs from

FIGURE 6-8

```
procedure negate ( a : inout word32 ) is
    variable carry_in : bit := '1';
    variable carry_out : bit;
begin
    a := not a;
    for index in a'reverse_range loop
        carry_out := a(index) and carry_in;
        a(index) := a(index) xor carry_in;
        carry_in := carry_out;
    end loop;
end procedure negate;
```

A procedure to negate an integer represented by a bit vector.

the parameter **result** in the **addu** procedure of the previous example.) We might include the following call to this procedure in a model:

variable op1 : word32;

. . .

negate (op1);

This uses the value of **op1** to initialize the formal parameter **a**. The procedure body is then executed, updating **a**, and when it returns, the final value of **a** is copied back into **op1**.

Signal Parameters

The third class of object that we can specify for formal parameters is *signal*, which indicates that the algorithm performed by the procedure involves a signal passed by the caller. A signal parameter can be of any of the modes **in**, **out** or **inout**. The way that signal parameters work is somewhat different from constant and variable parameters, so it is worth spending a bit of time understanding them.

When a caller passes a signal as a parameter of mode **in**, instead of passing the value of the signal, it passes the signal object itself. Any reference to the formal parameter within the procedure is exactly like a reference to the actual signal itself. A consequence of this is that if the procedure executes a wait statement, the signal value may be different after the wait statement completes and the procedure resumes. This behavior differs from that of constant parameters of mode **in**, which have the same value for the whole of the procedure.

EXAMPLE

Suppose we wish to model the receiver part of a network interface. It receives fixed-length packets of data on the signal **rx_data**. The data is synchronized with changes, from '0' to '1', of the clock signal **rx_clock**. Figure 6-9 is an outline of part of the model.

FIGURE 6-9

```
architecture behavioral of receiver is
    . . .        – – type declarations, etc
    signal recovered_data : bit;
    signal recovered_clock : bit;
    . . .
    procedure receive_packet ( signal rx_data : in bit;
                               signal rx_clock : in bit;
                               data_buffer : out packet_array ) is
    begin
        for index in packet_index_range loop
            wait until rx_clock = '1';
            data_buffer(index) := rx_data;
        end loop;
    end procedure receive_packet;
```

```
begin
    packet_assembler : process is
        variable packet : packet_array;
    begin
        . . .
        receive_packet ( recovered_data, recovered_clock, packet );
        . . .
    end process packet_assembler;
    . . .
end architecture behavioral;
```

An outline of a model of a network receiver, including a procedure with signal parameters of mode **in**.

During execution of the model, the process **packet_assembler** calls the procedure **receive_packet**, passing the signals **recovered_data** and **recovered_clock** as actual parameters. We can think of the procedure as executing "on behalf of" the process. When it reaches the wait statement, it is really the calling process that suspends. The wait statement mentions **rx_clock**, and since this stands for **recovered_clock**, the process is sensitive to changes on **recovered_clock** while it is suspended. Each time it resumes, it reads the current value of **rx_data** (which represents the actual signal **recovered_data**) and stores it in an element of the array parameter **data_buffer**.

Now let's look at signal parameters of mode **out**. In this case, the caller must name a signal as the actual parameter, and the procedure is passed a reference to the driver for the signal. When the procedure performs a signal assignment statement on the formal parameter, the transactions are scheduled on the driver for the actual signal parameter. In Chapter 5, we said that a process that contains a signal assignment statement contains a driver for the target signal, and that an ordinary signal may only have one driver. When such a signal is passed as an actual **out** mode parameter, there is still only the one driver. We can think of the signal assignments within the procedure as being performed on behalf of the process that calls the procedure.

EXAMPLE

Figure 6-10 is an outline of an architecture body for a signal generator. The procedure **generate_pulse_train** has **in** mode constant parameters that specify the characteristics of a pulse train and an **out** mode signal parameter on which it generates the required pulse train. The process **raw_signal_generator** calls the procedure, supplying **raw_signal** as the actual signal parameter for **s**. A reference to the driver for **raw_signal** is passed to the procedure, and transactions are generated on it.

An incidental point to note is the way we have specified the actual value for the **separation** parameter in the procedure call. This ensures that the sum of the **width** and **separation** values is exactly equal to **period**, even if **period** is not an even multiple of the time resolution limit. This illustrates an approach sometimes called "defensive programming," in which we try to ensure that the model works correctly in all possible circumstances.

FIGURE 6-10

```
library ieee;  use ieee.std_logic_1164.all;
architecture top_level of signal_generator is
    signal raw_signal : std_ulogic;
    . . .
    procedure generate_pulse_train ( width, separation : in delay_length;
                                     number : in natural;
                                     signal s : out std_ulogic ) is
    begin
        for count in 1 to number loop
            s <= '1', '0' after width;
            wait for width + separation;
        end loop;
    end procedure generate_pulse_train;
begin
    raw_signal_generator : process is
    begin
        . . .
        generate_pulse_train ( width => period / 2,
                               separation => period – period / 2,
                               number => pulse_count,
                               s => raw_signal );

        . . .
    end process raw_signal_generator;

    . . .
end architecture top_level;
```

An outline of a model for a signal generator, including a pulse generator procedure with an **out** *mode signal parameter.*

As with variable-class parameters, we can also have a signal-class parameter of mode **inout**. When the procedure is called, both the signal and a reference to its driver are passed to the procedure. The statements within it can read the signal value, include it in sensitivity lists in wait statements, query its attributes and schedule transactions using signal assignment statements.

A final point to note about signal parameters relates to procedures declared immediately within an architecture body. The target of any signal assignment statements within such a procedure must be a signal parameter, rather than a direct reference to a signal declared in the enclosing architecture body. The reason for this restriction is that the procedure may be called by more than one process within the architecture body. Each process that performs assignments on a signal has a driver for the signal. Without the restriction, we would not be able to tell easily by looking at the model where the drivers for the signal were located. The restriction makes the model more comprehensible and hence easier to maintain.

Default Values

The one remaining part of a procedure parameter list that we have yet to discuss is the optional default value expression, shown in the syntax rule on page 166. Note that we can only specify a default value for a formal parameter of mode **in**, and the parameter must be of the class constant or variable. If we include a default value in a parameter specification, we have the option of omitting an actual value when the procedure is called. We can either use the keyword **open** in place of an actual parameter value, or, if the actual value would be at the end of the parameter list, simply leave it out. If we omit an actual value, the default value is used instead.

EXAMPLE

Figure 6-11 is a procedure that increments an unsigned integer represented as a bit vector. The amount to increment by is specified by the second parameter, which has a default value of the bit-vector representation of 1.

FIGURE 6-11

```
procedure increment ( a : inout word32;  by : in word32 := X"0000_0001" ) is
    variable sum : word32;
    variable carry : bit := '0';
begin
    for index in a'reverse_range loop
        sum(index) := a(index) xor by(index) xor carry;
        carry := ( a(index) and by(index) ) or ( carry and ( a(index) xor by(index) ) );
    end loop;
    a := sum;
end procedure increment;
```

A procedure to increment a bit vector representing an unsigned integer.

If we have a variable count declared to be of type word32, we can call the procedure to increment it by 4, as follows:

    increment(count, X"0000_0004");

If we want to increment the variable by 1, we can make use of the default value for the second parameter and call the procedure without specifying an actual value to increment by, as follows:

    increment(count);

This call is equivalent to

    increment(count, by => **open**);

Unconstrained Array Parameters

In Chapter 4 we described unconstrained array types, in which the index range of the array was left unspecified using the "box" ("<>") notation. For such a type, we constrain the index bounds when we create an object, such as a variable or a signal. Another use of an unconstrained array type is as the type of a formal parameter to procedure. This use allows us to write a procedure in a general way, so that it can operate on array values of any size or with any range of index values. When we call the procedure and provide a constrained array as the actual parameter, the index bounds of the actual array are used as the bounds of the formal array parameter. Let us look at an example to show how unconstrained array parameters work.

EXAMPLE

Figure 6-12 is a procedure that finds the index of the first bit set to '1' in a bit vector. The formal parameter v is of type **bit_vector**, which is an unconstrained array type. Note that in writing this procedure, we do not explicitly refer to the index bounds of the formal parameter v, since they are not known. Instead, we use the 'range attribute.

IGURE 6-12

```
procedure find_first_set ( v : in bit_vector;
                          found : out boolean;
                          first_set_index : out natural ) is
begin
    for index in v'range loop
        if v(index) = '1' then
            found := true;
            first_set_index := index;
            return;
        end if;
    end loop;
    found := false;
end procedure find_first_set;
```

A procedure to find the first set bit in a bit vector.

When the procedure is executed, the formal parameters stand for the actual parameters provided by the caller. So if we call this procedure as follows:

```
variable int_req : bit_vector (7 downto 0);
variable top_priority : natural;
variable int_pending : boolean;
    . . .
find_first_set ( int_req, int_pending, top_priority );
```

v'range returns the range 7 **downto** 0, which is used to ensure that the loop parameter **index** iterates over the correct index values for v. If we make a different call:

```
variable free_block_map : bit_vector(0 to block_count–1);
variable first_free_block : natural;
variable free_block_found : boolean;
...
find_first_set ( free_block_map, free_block_found, first_free_block );
```

v'range returns the index range of the array free_block_map, since that is the actual parameter corresponding to v.

When we have formal parameters that are of array types, either constrained or unconstrained, we can use any of the array attributes mentioned in Chapter 4 to refer to the index bounds and range of the actual parameters. We can use the attribute values to define new local constants or variables whose index bounds and ranges depend on those of the parameters. The local objects are created anew each time the procedure is called.

EXAMPLE

The procedure in Figure 6-13 has two bit-vector parameters, which it assumes represent signed integer values in two's-complement form. It performs an arithmetic comparison of the numbers. It operates by taking temporary copies of each of the bit-vector parameters, inverting the sign bits and performing a lexical comparison using the built-in "<" operator. This is equivalent to an arithmetic comparison of the original numbers. Note that the temporary variables are declared to be of the same size as the parameters by using the 'range attribute, and the sign bits (the leftmost bits) are indexed using the 'left attribute.

FIGURE 6-13

```
procedure bv_lt ( bv1, bv2 : in bit_vector;  result : out boolean ) is
    variable tmp1 : bit_vector(bv1'range) := bv1;
    variable tmp2 : bit_vector(bv2'range) := bv2;
begin
    tmp1(tmp1'left) := not tmp1(tmp1'left);
    tmp2(tmp2'left) := not tmp2(tmp2'left);
    result :=  tmp1 < tmp2;
end procedure bv_lt;
```

A procedure to compare two bit vectors representing two's-complement signed integers.

Summary of Procedure Parameters

Let us now summarize all that we have seen in specifying and using parameters for procedures. The syntax rule on page 166 shows that we can specify five aspects of each formal parameter. First, we may specify the class of object, which determines how the formal parameter appears within the procedure, namely, as a constant, a variable or a signal. Second, we give a name to the formal parameter so that it can be referred to in the procedure body. Third, we may specify the mode, **in**, **out** or **inout**, which determines the direction in which information is passed between the caller and the procedure, and hence whether the procedure can read or assign to the formal parameter. Fourth, we must specify the type or subtype of the formal parameter, which restricts the type of actual parameters that can be provided by the caller. This is important as a means of preventing inadvertent misuse of the procedure. Fifth, we may include a default value, giving a value to be used if the caller does not provide an actual parameter. These five aspects clearly define the interface between the procedure and its callers, allowing us to partition a complex behavioral model into sections and concentrate on each section without being distracted by other details.

Once we have encapsulated some operations in a procedure, we can then call that procedure from different parts of a model, providing actual parameters to specialize the operation at each call. The syntax rule for a procedure call is

procedure_call_statement ⟸
    [label :] *procedure*_name [(*parameter*_association_list)] ;

This is a sequential statement, and so it may be used in a process or inside of another subprogram body. If the procedure has formal parameters, the call can specify actual parameters to associate with the formal parameters. The actual associated with a constant-class formal is the value of an expression. The actual associated with a variable-class formal must be a variable, and the actual associated with a signal-class formal must be a signal. The simplified syntax rule for the parameter association list is

*parameter*_association_list ⟸
    ([*parameter*_name =>]
        expression ‖ *signal*_name ‖ *variable*_name ‖ **open**) { , ... }

This is in fact the same syntax rule that applies to port maps in component instantiations, seen in Chapter 5. Most of what we said there also applies to procedure parameter association lists. For example, we can use positional association in the procedure call by providing one actual parameter for each formal parameter in the order listed in the procedure declaration. Alternatively, we can use named association by identifying explicitly which formal corresponds to which actual parameter in the call. In this case, the parameters can be in any order. Also, we can use a mix of positional and named association, provided all of the positional parameters come first in the call.

EXAMPLE
___

Suppose we have a procedure declared as

procedure p (f1 : **in** t1; f2 : **in** t2; f3 : **out** t3; f4 : **in** t4 := v4) **is**
begin
    . . .
end procedure p;

We could call this procedure, providing actual parameters in a number of ways, including

```
p ( val1, val2, var3, val4 );
p ( f1 => val1, f2 => val2, f4 => val4, f3 => var3 );
p ( val1, val2, f4 => open, f3 => var3 );
p ( val1, val2, var3 );
```

6.3 Concurrent Procedure Call Statements

In Chapter 5 we saw that VHDL provides concurrent signal assignment statements and concurrent assertions as shorthand notations for commonly used kinds of processes. Now that we have looked at procedures and procedure call statements, we can introduce another shorthand notation, the *concurrent procedure call statement*. As its name implies, it is short for a process whose body contains a sequential procedure call statement. The syntax rule is

```
concurrent_procedure_call_statement ⇐
    [ label : ] procedure_name [ ( parameter_association_list ) ] ;
```

This looks identical to an ordinary sequential procedure call, but the difference is that it appears as a concurrent statement, rather than as a sequential statement. A concurrent procedure call is exactly equivalent to a process that contains a sequential procedure call to the same procedure with the same actual parameters. For example, a concurrent procedure call of the form

```
call_proc : p ( s1, s2, val1 );
```

where s1 and s2 are signals and val1 is a constant, is equivalent to the process

```
call_proc : process is
begin
    p ( s1, s2, val1 );
    wait on s1, s2;
end process call_proc;
```

This also shows that the equivalent process contains a wait statement, whose sensitivity clause includes the signals mentioned in the actual parameter list. This is useful, since it results in the procedure being called again whenever the signal values change. Note that only signals associated with **in** mode or **inout** mode parameters are included in the sensitivity list. It would not make sense to include signals associated with **out** mode parameters, since the procedure never reads them but only assigns to them.

EXAMPLE

We can write a procedure that checks setup timing of a data signal with respect to a clock signal, as shown in Figure 6-14. When the procedure is called, it tests to see if there is a rising edge on the clock signal, and if so, checks that the data signal has not changed within the setup time interval. We can invoke this procedure using a concurrent procedure call, for example:

```
check_ready_setup : check_setup ( data => ready, clock => phi2,
                                   Tsu => Tsu_rdy_clk );
```

FIGURE 6-14

```
procedure check_setup ( signal data, clock : in bit;
                        constant Tsu : in time ) is
begin
    if clock'event and clock = '1' then
        assert data'last_event >= Tsu
            report "setup time violation" severity error;
    end if;
end procedure check_setup;
```

A procedure to check setup timing of a data signal.

The procedure is called whenever either of the signals in the actual parameter list, **ready** or **phi2**, changes value. When the procedure returns, the concurrent procedure call statement suspends until the next event on either signal. The advantage of using a concurrent procedure call like this is twofold. First, we can write a suite of commonly used checking procedures and reuse them whenever we need to include a check in a model. This is potentially a great improvement in productivity. Second, the statement that invokes the check is more compact and readily understandable than the equivalent process written in-line.

Another point to note about concurrent procedure calls is that if there are no signals associated with **in** mode or **inout** mode parameters, the wait statement in the equivalent process does not have a sensitivity clause. If the procedure ever returns, the process suspends indefinitely. This may be useful if we only want the procedure to be called once at startup time. On the other hand, we may write the procedure so that it never returns. If we include wait statements within a loop in the procedure, it behaves somewhat like a process itself. The advantage of this is that we can declare a procedure that performs some commonly needed behavior and then invoke one or more instances of it using concurrent procedure call statements.

EXAMPLE

The procedure in Figure 6-15 generates a periodic clock waveform on a signal passed as a parameter. The **in** mode constant parameters specify the shape of a clock waveform. The procedure waits for the initial phase delay, then loops indefinitely, scheduling a new rising and falling transition on the clock signal parameter on each iteration. It never returns to its caller. We can use this procedure to generate a two-phase non-overlapping pair of clock signals, as follows:

```
signal phi1, phi2 : std_ulogic := '0';
. . .
gen_phi1 : generate_clock ( phi1, Tperiod => 50 ns,
                            Tpulse => 20 ns, Tphase => 0 ns );
gen_phi2 : generate_clock ( phi2, Tperiod => 50 ns,
                            Tpulse => 20 ns, Tphase => 25 ns );
```

FIGURE 6-15

```
procedure generate_clock ( signal clk : out std_ulogic;
                           constant Tperiod, Tpulse, Tphase : in time ) is
begin
    wait for Tphase;
    loop
        clk <= '1', '0' after Tpulse;
        wait for Tperiod;
    end loop;
end procedure generate_clock;
```

A procedure that generates a clock waveform on a signal.

Each of these calls represents a process that calls the procedure, which then executes the clock generation loop on behalf of its parent process. The advantage of this approach is that we only had to write the loop once in a general-purpose procedure. Also, we have made the model more compact and understandable.

6.4   Functions

Let us now turn our attention to the second kind of subprogram in VHDL: *functions*. We can think of a function as a generalization of expressions. The expressions that we described in Chapter 2 combined values with operators to produce new values. A function is a way of defining a new operation that can be used in expressions. We define how the new operation works by writing a collection of sequential statements that calculate the result. The syntax rule for a function declaration is very similar to that for a procedure declaration:

subprogram_body ⇐
    ⟦ **pure** ∥ **impure** ⟧
    function identifier
             ⟦ (*parameter*_interface_list) ⟧ **return** type_mark **is**
      { subprogram_declarative_item }
    begin
      { sequential_statement }
    end ⟦ **function** ⟧ ⟦ identifier ⟧ ;

The identifier in the declaration names the function. It may be repeated at the end of the declaration. Unlike a procedure subprogram, a function calculates and returns a result that can be used in an expression. The function declaration specifies the type of the result after the keyword **return**. The parameter list of a function takes the same form as that for a procedure, with two restrictions. First, the parameters of a function may not be of the class variable. If the class is not explicitly mentioned, it is assumed to be constant. Second, the mode of each parameter must be **in**. If the mode is not explicitly specified, it is assumed to be **in**. We come to the reasons for these restrictions in a moment. Like a procedure, a function can declare local items in its declarative part for use in the statements in the function body.

A function passes the result of its computation back to its caller using a return statement, given by the syntax rule

return_statement ⇐ ⟦ label : ⟧ **return** expression ;

The optional label allows us to identify the return statement. The form described by this syntax rule differs from the return statement in a procedure subprogram in that it includes an expression to provide the function result. Furthermore, a function must include at least one return statement of this form, and possibly more. The first to be executed causes the function to complete and return its result to the caller. A function cannot simply run into the end of the function body, since to do so would not provide a way of specifying a result to pass back to the caller.

A function call looks exactly like a procedure call. The syntax rule is

function_call ⇐ *function*_name ⟦ (*parameter*_association_list) ⟧

The difference is that a function call is part of an expression, rather than being a sequential statement on its own, like a procedure call.

EXAMPLE

Figure 6-16 is a simple function that calculates whether a value is within given bounds and returns a result limited to those bounds. A call to this function might be included in a variable assignment statement, as follows:

```
new_temperature := limit ( current_temperature + increment,
                           10, 100 );
```

In this statement, the expression on the right-hand side of the assignment consists of just the function call, and the result returned is assigned to the variable **new_temperature**. However, we might also use the result of a function call in further computation, for example:

```
new_motor_speed := old_motor_speed
                   + scale_factor * limit ( error, −10, +10 );
```

FIGURE 6-16

```
function limit ( value, min, max : integer ) return integer is
begin
    if value > max then
        return max;
    elsif value < min then
        return min;
    else
        return value;
    end if;
end function limit;
```

A function to limit a value to specified bounds.

EXAMPLE

The function in Figure 6-17 determines the number represented in binary by a bit-vector value. The algorithm scans the bit vector from the most-significant end. For each bit, it multiplies the previously accumulated value by two and then adds in the integer value of the bit. The accumulated value is then used as the result of the function, passed back to the caller by the return statement.

FIGURE 6-17

```
function bv_to_natural ( bv : in bit_vector ) return natural is
    variable result : natural := 0;
begin
    for index in bv'range loop
        result := result * 2 + bit'pos(bv(index));
    end loop;
    return result;
end function bv_to_natural;
```

A function that converts the binary representation of an unsigned number to a numeric value.

As an example of using this function, consider a model for a read-only memory, which represents the stored data as an array of bit vectors, as follows:

```
type rom_array is array (natural range 0 to rom_size–1)
                        of bit_vector(0 to word_size–1);
variable rom_data : rom_array;
```

If the model has an address port that is a bit vector, we can use the function to convert the address to a natural value to index the ROM data array, as follows:

```
data <= rom_data ( bv_to_natural(address) ) after Taccess;
```

VHDL-87

The keyword **function** may not be included at the end of a function declaration in VHDL-87. Return statements may not be labeled in VHDL-87.

Functional Modeling

In Chapter 5 we looked at concurrent signal assignment statements for functional modeling of designs. We can use functions in VHDL to help us write functional models more expressively by defining a function that encapsulates the data transformation to be performed and then calling the function in a concurrent signal assignment statement. For example, given a declaration of a function to add two bit vectors:

```
function bv_add ( bv1, bv2 : in bit_vector ) return bit_vector is
begin
    . . .
end function bv_add;
```

and signals declared in an architecture body:

> **signal** source1, source2, sum : bit_vector(0 **to** 31);

we can write a concurrent signal assignment statement as follows:

> adder : sum <= bv_add(source1, source2) **after** T_delay_adder;

Pure and Impure Functions

Let us now return to the reason for the restrictions on the class and mode of function formal parameters stated above. These restrictions are in keeping with our idea that a function is a generalized form of operator. If we pass the same values to an operator, such as the addition operator, in different expressions, we expect the operator to return the same result each time. By restricting the formal parameters of a function in the way described above, we go part of the way to ensuring the same property for function calls. One additional restriction we need to make is that the function may not refer to any variables or signals declared by its parents, that is, by any process, subprogram or architecture body in which the function declaration is nested. Otherwise the variables or signals might change values between calls to the function, thus influencing the result of the function. We call a function that makes no such reference a *pure* function. We can explicitly declare a function to be pure by including the keyword **pure** in its definition, as shown by the syntax rule on page 179. If we leave it out, the function is assumed to be pure. Both of the above examples of function declarations are pure functions.

On the other hand, we may deliberately relax the restriction about a function referencing its parents' variables or signals by including the keyword **impure** in the function declaration. This is a warning to any caller of the function that it might produce different results on different calls, even when passed the same actual parameter values.

EXAMPLE

Many network protocols require a sequence number in the packet header so that they can handle packets getting out of order during transmission. We can use an impure function to generate sequence numbers when creating packets in a behavioral model of a network interface. Figure 6-18 is an outline of a process that represents the output side of the network interface.

In this model, the process has a variable **next_seq_number**, used by the function **generate_seq_number** to determine the return value each time it is called. The function has the side effect of incrementing this variable, thus changing the value to be returned on the next call. Because of the reference to the variable in the function's parent, the function must be declared to be impure. The advantage of writing the function this way lies in the expressive power of its call. The function call is simply part of an expression, in this case yielding an element in a record aggregate of type **pkt_header**. Writing it this way makes the process body more compact and easily understandable.

FIGURE 6-18

```
network_driver : process is
    constant seq_modulo : natural := 2**5;
    subtype seq_number is natural range 0 to seq_modulo-1;
    variable next_seq_number : seq_number := 0;
    . . .
    impure function generate_seq_number return seq_number is
        variable number : seq_number;
    begin
        number := next_seq_number;
        next_seq_number := (next_seq_number + 1) mod seq_modulo;
        return number;
    end function generate_seq_number;
begin  - - network_driver
    . . .
    new_header := pkt_header'( dest => target_host_id,
                               src => my_host_id,
                               pkt_type => control_pkt,
                               seq => generate_seq_number );
    . . .
end process network_driver;
```

An outline of a network driver process, including an impure function to calculate sequence numbers for network packets.

The Function Now

VHDL provides a predefined function, now, that returns the current simulation time when it is called. It is defined as

impure function now **return** delay_length;

It is defined to be an impure function because it returns a different value when called at different times during the course of a simulation. Recall that the type delay_length is a predefined subtype of the physical type time, constrained to non-negative time values. The function now is often used to check that the inputs to a model obey the required timing constraints.

EXAMPLE

Figure 6-19 is a process that checks the clock and data inputs of an edge-triggered flipflop for adherence to the minimum hold time constraint, Thold_d_clk. When the clock signal changes to '1', the process saves the current simulation time in the variable last_clk_edge_time. When the data input changes, the process tests whether the current simulation time has advanced beyond the time of the last clock edge by at least the minimum hold time, and reports an error if it has not.

FIGURE 6-19

```
hold_time_checker : process ( clk, d ) is
    variable last_clk_edge_time : time := 0 fs;
begin
    if clk'event and clk = '1' then
        last_clk_edge_time := now;
    end if;
    if d'event then
        assert now – last_clk_edge_time >= Thold_d_clk
            report "hold time violation";
    end if;
end process hold_time_checker;
```

A process that checks for data hold time after clock rising edges for an edge-triggered flipflop.

VHDL-87

The function now returns a value of type time in VHDL-87, since the subtype delay_length is not predefined in VHDL-87.

6.5 Overloading

When we are writing subprograms, it is a good idea to choose names for our subprograms that indicate what operations they perform, to make it easier for a reader to understand our models. This raises the question of how to name two subprograms that perform the same kind of operation but on parameters of different types. For example, we might wish to write two procedures to increment variables holding numeric values, but in some cases the values are represented as type integer, and in other cases they are represented using type bit_vector. Ideally, since both procedures perform the same operation, we would like to give them the same name, such as increment. But if we did that, would we be able to tell them apart when we wanted to call them? Recall that VHDL strictly enforces the type rules, so we have to refer to the right procedure depending on the type of the variable we wish to increment.

Fortunately, VHDL allows us to define subprograms in this way, using a technique called *overloading* of subprogram names. We can define two distinct subprograms with the same name but with different numbers or types of formal parameters. When we call one of them, the number and types of the actual parameters we supply in the call are used to determine which subprogram to invoke. It is the context of the call that determines how to resolve the apparent ambiguity. We have already seen overloading applied to identifiers used as literals in enumeration types (see Chapter 2). We saw that if two enumeration types included the same identifier, the context of use in a model is used to determine which type is meant.

The precise rules used to disambiguate a subprogram call when the subprogram name is overloaded are quite complex, so we will not enumerate them all here. Fortunately, they are sufficiently complete to sort out most situations that arise in practice.

Instead, we look at some examples to show how overloading of procedures and functions works in straightforward cases. First, here are some procedure outlines for the increment operation described above:

procedure increment (a : **inout** integer;   n : **in** integer := 1) **is** . . .

procedure increment (a : **inout** bit_vector; n : **in** bit_vector := B"1") **is** . . .

procedure increment (a : **inout** bit_vector; n : **in** integer := 1) **is** . . .

Suppose we also have some variables declared as follows:

variable count_int : integer := 2;
variable count_bv : bit_vector (15 **downto** 0) := X"0002";

If we write a procedure call using count_int as the first actual parameter, it is clear that we are referring to the first procedure, since it is the only one whose first formal parameter is an integer. Both of the following calls can be disambiguated in this way:

increment (count_int, 2);
increment (count_int);

Similarly, both of the next two calls can be sorted out:

increment (count_bv, X"0002");
increment (count_bv, 1);

The first call refers to the second procedure, since the actual parameters are both bit vectors. Similarly, the second call refers to the third procedure, since the actual parameters are a bit vector and an integer. Problems arise, however, if we try to make a call as follows:

increment (count_bv);

This could equally well be a call to either the second or the third procedure, both of which have default values for the second formal parameter. Since it is not possible to determine which procedure is meant, a VHDL analyzer rejects such a call as an error.

Overloading Operator Symbols

When we introduced function subprograms in Section 6.4, we described them as a generalization of operators used in expressions, such as "+", "–", **and**, **or** and so on. Looking at this the other way around, we could say that the predefined operators are specialized functions, with a convenient notation for calling them. In fact, this is exactly what they are. Furthermore, since each of the operators can be applied to values of various types, we see that the functions they represent are overloaded, so the types of the operands determine the particular version of each operator used in an expression.

Given that we can define our own types in VHDL, it would be convenient if we could extend the predefined operators to work with these types. For example, if we are using bit vectors to model integers using two's-complement notation, we would like to use the addition operator to add two bit vectors in this form. Fortunately, VHDL provides a way for us to define new functions using the operator symbols as names. Our bit-vector addition function can be declared as

```
function "+" ( left, right : in bit_vector ) return bit_vector is
begin
    . . .
end function "+";
```

We can then call this function using the infix "+" operator with bit-vector operands, for example:

```
variable addr_reg : bit_vector(31 downto 0);
. . .
addr_reg := addr_reg + X"0000_0004";
```

Operators denoted by reserved words can be overloaded in the same way. For example, we can declare a bit-vector absolute-value function as

```
function "abs" ( right : in bit_vector ) return bit_vector is
begin
    . . .
end function "abs";
```

We can use this operator with a bit-vector operand, for example:

```
variable accumulator : bit_vector(31 downto 0);
. . .
accumulator := abs accumulator;
```

We can overload any of the operator symbols shown in Figure 2-5 on pages 50–51. One important point to note, however, is that overloaded versions of the logical operators **and**, **nand**, **or** and **nor** are not evaluated in the short-circuit manner described in Chapter 2. For any type of operands other than bit and boolean, both operands are evaluated first, then passed to the function.

EXAMPLE

The std_logic_1164 package defines functions for logical operators applied to values of type std_ulogic and std_ulogic_vector. We can use them in functional models to write Boolean equations that represent the behavior of a design. For example, Figure 6-20 describes a block of logic that controls an input/output register in a microcontroller system. The architecture body describes the behavior in terms of Boolean equations. Its concurrent signal assignment statements use the logical operators **and** and **not**, referring to the overloaded functions defined in the std_logic_1164 package.

FIGURE 6-20

```
library ieee;  use ieee.std_logic_1164.all;
entity reg_ctrl is
    port ( reg_addr_decoded, rd, wr, io_en, cpu_clk : in std_ulogic;
            reg_rd, reg_wr : out std_ulogic );
end entity reg_ctrl;
```

```
architecture bool_eqn of reg_ctrl is
begin
        rd_ctrl : reg_rd <= reg_addr_decoded and rd and io_en;
        rw_ctrl : reg_wr <= reg_addr_decoded and wr and io_en
                                                and not cpu_clk;
end architecture bool_eqn;
```

An entity and architecture body for a logic block that controls operation of a register.

VHDL-87

Since VHDL-87 does not provide the shift operators **sll**, **srl**, **sla**, **sra**, **rol**, and **ror** and the logical operator **xnor**, they cannot be used as operator symbols.

6.6 Visibility of Declarations

The last topic we need to discuss in relation to subprograms is the use of names declared within a model. We have seen that names of types, constants, variables and other items defined in a subprogram can be used in that subprogram. Also, in the case of procedures and impure functions, names declared in an enclosing process, subprogram or architecture body can also be used. The question we must answer is: What are the limits of use of each name?

To answer this question, we introduce the idea of the *visibility* of a declaration, which is the region of the text of a model in which it is possible to refer to the declared name. We have seen that architecture bodies, processes and subprograms are each divided into two parts: a declarative part and a body of statements. A name declared in a declarative part is visible from the end of the declaration itself down to the end of the corresponding statement part. Within this area we can refer to the declared name. Before the declaration, within it and beyond the end of the statement part, we cannot refer to the name because it is not visible.

EXAMPLE

Figure 6-21 shows an outline of an architecture body of a model. It contains a number of declarations, including some procedure declarations. The visibility of each of the declarations is indicated. The first item to be declared is the type t; its visibility extends to the end of the architecture body. Thus it can be referred in other declarations, such as the variable declarations. The second declaration is the signal s; its visibility likewise extends to the end of the architecture body. So the assignment within procedure p1 is valid. The third and final declaration in the declarative part of the architecture body is that of the procedure p1, whose visibility extends to the end of the architecture body, allowing it to be called in either of the processes. It includes a local variable, v1, whose visibility extends only to the end of p1. This means it can be referred to in p1, as shown in the signal assignment statement, but neither process can refer to it.

FIGURE 6-21

```
architecture arch of ent is
    type t is . . .;                                                                t
    signal s : t;                                                             s
    procedure p1 ( . . . ) is                                    p1
        variable v1 : t;                          v1
    begin
        v1 := s;
    end procedure p1;
begin  – – arch
    proc1 : process is
        variable v2 : t;                                   v2
        procedure p2 ( . . . ) is              p2
            variable v3 : t;              v3
        begin
            p1 ( v2, v3, . . . );
        end procedure p2;
    begin  – – proc1
        p2 ( v2, . . . );
    end process proc1;
    proc2 : process is
        . . .
    begin  – – proc2
        p1 ( . . . );
    end process proc2;
end architecture arch;
```

An outline of an architecture body, showing the visibility of declared names within it.

In the statement part of the architecture body, we have two process state-ments, **proc1** and **proc2**. The first includes a local variable declaration, **v2**, whose visibility extends to the end of the process body. Hence we can refer to **v2** in the process body and in the procedure **p2** declared within the process. The visibility of **p2** likewise extends to the end of the body of **proc1**, allowing us to call **p2** within **proc1**. The procedure **p2** includes a local variable declaration, **v3**, whose visibility extends to the end of the statement part of **p2**. Hence we can refer to **v3** in the statement part of **p2**. However, we cannot refer to **v3** in the statement part of **proc1**, since it is not visible in that part of the model.

Finally, we come to the second process, **proc2**. The only items we can refer to here are those declared in the architecture body declarative part, namely, **t**, **s** and **p1**. We cannot call the procedure **p2** within **proc2**, since it is local to **proc1**.

One point we mentioned earlier about subprograms but did not go into in detail was that we can include nested subprogram declarations within the declarative part of a subprogram. This means we can have local procedures and functions within a procedure or a function. In such cases, the simple rule for the visibility of a declaration still applies, so any items declared within an outer procedure before the declaration of a nested procedure can be referred to inside the nested procedure.

EXAMPLE

Figure 6-22 is an outline of an architecture of a cache memory for a computer system. The entity interface includes ports named mem_addr, mem_ready, mem_ack and mem_data_in. The process behavior contains a procedure, read_block, which reads a block of data from main memory on a cache miss. It has the local variables memory_address_reg and memory_data_reg. Nested inside of this procedure is another procedure, read_memory_word, which reads a single word of data from memory. It uses the value placed in memory_address_reg by the outer procedure and leaves the data read from memory in memory_data_reg.

FIGURE 6-22

```
architecture behavioral of cache is
begin
    behavior : process is

        . . .

        procedure read_block ( start_address : natural;  entry : out cache_block ) is
            variable memory_address_reg : natural;
            variable memory_data_reg : word;

            procedure read_memory_word is
            begin
                mem_addr <= memory_address_reg;  mem_read <= '1';
                wait until mem_ack = '1';
                memory_data_reg := mem_data_in;  mem_read <= '0';
                wait until mem_ack = '0';
            end procedure read_memory_word;

        begin  - - read_block
            for offset in 0 to block_size - 1 loop
                memory_address_reg := start_address + offset;
                read_memory_word;
                entry(offset) := memory_data_reg;
            end loop;
        end procedure read_block;
    begin  - - behavior
        . . .
        read_block ( miss_base_address, data_store(entry_index) );
        . . .
    end process behavior;
end architecture behavioral;
```

An outline of a behavioral architecture of a cache memory.

Now let us consider a model in which we have one subprogram nested inside of another, and each declares an item with the same name as the other, as shown in Figure 6-23. Here, the first variable v is visible within all of the procedure p2 and the statement body of p1. However, because p2 declares its own local variable called v, the variable belonging to p1 is not *directly visible* where p2's v is visible. We say the inner variable declaration *hides* the outer declaration, since it declares the same name. Hence the addition within p2 applies to the local variable v of p2 and does not affect the variable v of p1. If we need to refer to an item that is visible but hidden, we can use a selected name. For example, within p2 in Figure 6-23, we can use the name p1.v to refer to the variable v declared in p1. Although the outer declaration is not directly visible, it is *visible by selection*.

FIGURE 6-23

```
procedure p1 is
     variable v : integer;                    v
     procedure p2 is
          variable v : integer;        v
     begin  --p2
          . . .
          v := v + 1;
          . . .
     end procedure p2;
begin  --p1
     . . .
     v := 2 * v;
     . . .
end procedure p1;
```

Nested procedures showing hiding of names. The declaration of v in p2 hides the variable v declared in p1.

The idea of hiding is not restricted to variable declarations within nested procedures. Indeed, it applies in any case where we have one declarative part nested within another, and an item is declared with the same name in each declarative part in such a way that the rules for resolving overloaded names are unable to distinguish between them. The advantage of having inner declarations hide outer declarations, as opposed to the alternative of simply disallowing an inner declaration with the same name, is that it allows us to write local procedures and processes without having to know the names of all items declared at outer levels. This is certainly beneficial when writing large models. In practice, if we are reading a model and need to check the use of a name in a statement against its declaration, we only need to look at successively enclosing declarative parts until we find a declaration of the name, and that is the declaration that applies.

Exercises

1. [❶ 6.2] Write parameter specifications for the following constant-class parameters:
   * an integer, operand1,
   * a bit vector, tag, indexed from 31 down to 16, and
   * a Boolean, trace, with default value false.

2. [❶ 6.2] Write parameter specifications for the following variable-class parameters:
   * a real number, average, used to pass data back from a procedure, and
   * a string, identifier, modified by a procedure.

3. [❶ 6.2] Write parameter specifications for the following signal-class parameters:
   * a bit signal, clk, to be assigned to by a procedure, and
   * an unconstrained standard-logic vector signal, data_in, whose value is to be read by a procedure.

4. [❶ 6.2] Given the following procedure declaration:

   ```
   procedure stimulate ( signal target : out bit_vector;
                         delay : in delay_length := 1 ns;
                         cycles : in natural := 1 ) is . . .
   ```

   write procedure calls using a signal s as the actual parameter for target and using the following values for the other parameters:
   * delay = 5 ns, cycles = 3,
   * delay = 10 ns, cycles = 1, and
   * delay = 1 ns, cycles = 15.

5. [❶ 6.3] Suppose we have a procedure declared as

   ```
   procedure shuffle_bytes (signal d_in : in std_ulogic_vector(0 to 15);
                            signal d_out : out std_ulogic_vector(0 to 15);
                            signal shuffle_control : in std_ulogic;
                            prop_delay : delay_length ) is . . .
   ```

   Write the equivalent process for the following concurrent procedure call:

   ```
   swapper : shuffle_bytes ( ext_data, int_data, swap_control, Tpd_swap );
   ```

6. [❶ 6.4] Suppose we have a function declared as

   ```
   function approx_log_2 ( a : in bit_vector ) return positive is . . .
   ```

   that calculates the minimum number of bits needed to represent a binary-encoded number. Write a variable assignment statement that calculates the minimum number of bits needed to represent the product of two numbers in the variables multiplicand and multiplier, as assigns the result to the variable product_size.

7. [❶ 6.4] Write an assertion statement that verifies that the current simulation time has not exceeded 20 ms.

8. [❶ 6.5] Given the declarations of the three procedures named **increment** and the variables **count_int** and **count_bv** shown on page 185, which of the three procedures, if any, is referred to by each of the following procedure calls?

    ```
    increment ( count_bv, –1 );
    increment ( count_int );
    increment ( count_int, B"1" );
    increment ( count_bv, 16#10# );
    ```

9. [❶ 6.6] Show the parts of the following model in which each of the declared items is visible:

    ```
    architecture behavioral of computer system is

        signal internal_data : bit_vector(31 downto 0);

        interpreter : process is

            variable opcode : bit_vector(5 downto 0);

            procedure do_write is
                variable aligned_address : natural;
            begin
                . . .
            end procedure do_write;

        begin
            . . .
        end process interpreter;

    end architecture behavioral;
    ```

10. [❷ 6.1] Write a procedure that calculates the sum of squares of elements of an array variable **deviations**. The elements are real numbers. Your procedure should store the result in a real variable **sum_of_squares**.

11. [❷ 6.1] Write a procedure that generates a 1 μs pulse every 20 μs on a signal **syn_clk**. When the signal **reset** changes to '1', the procedure should immediately set **syn_clk** to '0' and return.

12. [❷ 6.2] Write a procedure called **align_address** that aligns a binary encoded address in a bit-vector variable parameter. The procedure has a second parameter that indicates the alignment size. If the size is 1, the address is unchanged. If the size is 2, the address is rounded to a multiple of 2 by clearing the least-significant bit. If the size is 4, two bits are cleared, and if the size is 8, three bits are cleared. The default alignment size is 4.

13. [❷ 6.2/6.3] Write a procedure that checks the hold time of a data signal with respect to rising edges of a clock signal. Both signals are of the IEEE standard-logic type. The signals and the hold time are parameters of the procedure. The procedure is invoked by a concurrent procedure call.

14. [❷ 6.2/6.3] Write a procedure, to be invoked by a concurrent procedure call, that assigns successive natural numbers to a signal at regular intervals. The signal and the interval between numbers are parameters of the procedure.

15. [❷ 6.4] Write a function, **weaken**, that maps a standard-logic value to the same value, but with weak drive strength. Thus, '0' and 'L' are mapped to 'L', '1' and 'H'

are mapped to 'H', 'X' and 'W' are mapped to 'W' and all other values are unchanged.

16. [❷ 6.4] Write a function, returning a Boolean result, that tests whether a standard-logic signal currently has a valid edge. A valid edge is defined to be a transition from '0' or 'L' to '1' or 'H' or vice versa. Other transitions, such as 'X' to '1', are not valid.

17. [❷ 6.4] Write two functions, one to find the maximum value in an array of integers and the other to find the minimum value.

18. [❷ 6.5] Write overloaded versions of the logical operators to operate on integer operands. The operators should treat the value 0 as logical falsehood and any non-zero value as logical truth.

19. [❸ 6.2] Write a procedure called **scan_results** with an **in** mode bit-vector signal parameter **results**, and **out** mode variable parameters **majority_value** of type **bit**, **majority_count** of type **natural** and **tie** of type **boolean**. The procedure counts the occurrences of '0' and '1' values in **results**. It sets **majority_value** to the most frequently occurring value, **majority_count** to the number of occurrences and **tie** to true if there are an equal number of occurrences of '0' and '1'.

20. [❸ 6.2/6.3] Write a procedure that stimulates a bit-vector signal passed as a parameter. The procedure assigns to the signal a sequence of all possible bit-vector values. The first value is assigned to the signal immediately, then subsequent values are assigned at intervals specified by a second parameter. After the last value is assigned, the procedure returns.

21. [❸ 6.2/6.3] Write a passive procedure that checks that setup and hold times for a data signal with respect to rising edges of a clock signal are observed. The signals and the setup and hold times are parameters of the procedure. Include a concurrent procedure call to the procedure in the statement part of a D-flipflop entity.

22. [❸ 6.4] Write a function that calculates the cosine of a real number, using the series

$$\cos\theta = 1 - \frac{\theta^2}{2!} + \frac{\theta^4}{4!} - \frac{\theta^6}{6!} + \cdots$$

Next, write a second function that returns a cosine table of the following type:

type table **is array** (0 **to** 1023) **of** real;

Element i of the table has the value $\cos i\pi/2048$. Finally, develop a behavioral model of a cosine lookup ROM. The architecture body should include a constant of type **table**, initialized using a call to the second function.

7

Packages and Use Clauses

Packages in VHDL provide an important way of organizing the data and subprograms declared in a model. In this chapter, we describe the basics of packages and show how they may be used. We also look at one of the predefined packages, which includes all of the predefined types and operators available in VHDL.

7.1   Package Declarations

A VHDL package is simply a way of grouping a collection of related declarations that serve a common purpose. They might be a set of subprograms that provide operations on a particular type of data, or they might just be the set of declarations needed to model a particular design. The important thing is that they can be collected together into a separate design unit that can be worked on independently and reused in different parts of a model.

Another important aspect of packages is that they separate the external view of the items they declare from the implementation of those items. The external view is specified in a *package declaration*, whereas the implementation is defined in a separate *package body*. We will look at package declaration first and return to the package body shortly.

The syntax rule for writing a package declaration is

package_declaration ⇐
    package identifier **is**
        { package_declarative_item }
    end [**package**] [identifier] ;

The identifier provides a name for the package, which we can use elsewhere in a model to refer to the package. Inside the package declaration we write a collection of declarations, including type, subtype, constant, signal and subprogram declarations, as well as several other kinds of declarations that we see in later chapters. These are the declarations that are provided to the users of the package. The advantage of placing them in a package is that they do not clutter up other parts of a model, and they can be shared within and between models without having to rewrite them. Figure 7-1 is a simple example of a package declaration.

FIGURE 7-1
_____

```
package cpu_types is

    constant word_size : positive := 16;
    constant address_size : positive := 24;

    subtype word is bit_vector(word_size – 1 downto 0);
    subtype address is bit_vector(address_size – 1 downto 0);

    type status_value is ( halted, idle, fetch, mem_read, mem_write,
                           io_read, io_write, int_ack );

end package cpu_types;
```

_____

A package that declares some useful constants and types for a CPU model.

VHDL-87
_____

The keyword **package** may not be included at the end of a package declaration in VHDL-87.

_____

A package is another form of design unit, along with entity declarations and architecture bodies. It is separately analyzed and is placed into the working library as a library unit by the analyzer. From there, other library units can refer to an item declared in the package using the *selected name* of the item. The selected name is formed by writing the library name, then the package name and then the name of the item, all separated by dots; for example:

work.cpu_types.status_value

EXAMPLE

Suppose the **cpu_types** package, shown in Figure 7-1, has been analyzed and placed into the **work** library. We might make use of the declared items when modeling an address decoder to go with a CPU. The entity declaration and architecture body of the decoder are shown in Figure 7-2.

FIGURE 7-2

```
entity address_decoder is
    port ( addr : in work.cpu_types.address;
           status : in work.cpu_types.status_value;
           mem_sel, int_sel, io_sel : out bit );
end entity address_decoder;

- - - - - - - - - - - - - - - - - - - - - - - - - - - - - - - - - - - - -

architecture functional of address_decoder is
    constant mem_low : work.cpu_types.address := X"000000";
    constant mem_high : work.cpu_types.address := X"EFFFFF";
    constant io_low : work.cpu_types.address := X"F00000";
    constant io_high : work.cpu_types.address := X"FFFFFF";
begin
    mem_decoder :
        mem_sel <=
            '1' when ( work.cpu_types."="(status, work.cpu_types.fetch)
                        or work.cpu_types."="(status, work.cpu_types.mem_read)
                        or work.cpu_types."="(status, work.cpu_types.mem_write) )
                    and addr >= mem_low and addr <= mem_high else
            '0';
    int_decoder :
        int_sel <= '1' when work.cpu_types."="(status, work.cpu_types.int_ack) else
                    '0';
    io_decoder :
        io_sel <=
            '1' when ( work.cpu_types."="(status, work.cpu_types.io_read)
                        or work.cpu_types."="(status, work.cpu_types.io_write) )
                    and addr >= io_low and addr <= io_high else
            '0';
end architecture functional;
```

An entity and architecture body for an address decoder, using items declared in the cpu_types *package.*

Note that we have to use selected names to refer to the subtype **address**, the type **status_value**, the enumeration literals of **status_value** and the implicitly declared "=" operator, defined in the package **cpu_types**. This is because they are not directly visible within the entity declaration and architecture body. We will see later in this chapter how a use clause can help us avoid long selected names. If we needed to type-qualify the enumeration literals, we would use selected names for both the type name and the literal name; for example:

```
work.cpu_types.status_value'(work.cpu_types.fetch)
```

We have seen that a package, when analyzed, is placed into the working library. Items in the package can be accessed by other library units using selected names starting with **work**. However, if we are writing a package of generally useful declarations, we may wish to place them into a different library, such as a project library, where they can be accessed by other designers. Different VHDL tool suites provide different ways of specifying the library into which a library unit is placed. We must consult the documentation for a particular product to find out what to do. However, once the package has been included in a resource library, we can refer to items declared in it using selected names, starting with the resource library name. As an example, we might consider the IEEE standard-logic package, which must be placed in a resource library called **ieee**. We can refer to the types declared in that package, for example:

```
variable stored_state : ieee.std_logic_1164.std_ulogic;
```

One kind of declaration we can include in a package declaration is a signal declaration. This gives us a way of defining a signal, such as a master clock or reset signal, that is global to a whole design, instead of being restricted to a single architecture body. Any module that needs to refer to the global signal simply names it using the selected name as described above. This avoids the clutter of having to specify the signal as a port in each entity that uses it, making the model a little less complex. However, it does mean that a module can affect the overall behavior of a system by means other than through its ports, namely, by assigning to global signals. This effectively means that part of the module's interface is implicit, rather than being specified in the port map of the entity. As a matter of style, global signals declared in packages should be used sparingly, and their use should be clearly documented with comments in the model.

EXAMPLE

The package shown in Figure 7-3 declares two clock signals for use within an integrated circuit design for an input/output interface controller. The top-level architecture of the controller circuit is outlined in Figure 7-4. The instance of the **phase_locked_clock_gen** entity uses the **ref_clock** port of the circuit to generate the two-phase clock waveforms on the global clock signals. The architecture also includes an instance of an entity that sequences bus operations using the bus control signals and generates internal register control signals. The architecture body for the sequencer is outlined in Figure 7-5. It creates an instance of a register entity and connects the global clock signals to its clock input ports.

FIGURE 7-3

```
library ieee;  use ieee.std_logic_1164.all;
package clock_pkg is
    constant Tpw : delay_length := 4 ns;
    signal clock_phase1, clock_phase2 : std_ulogic;
end package clock_pkg;
```

A package that declares global clock signals.

FIGURE 7-4

```
library ieee;  use ieee.std_logic_1164.all;
entity io_controller is
    port ( ref_clock : in std_ulogic;  . . . );
end entity io_controller;
─ ─ ─ ─ ─ ─ ─ ─ ─ ─ ─ ─ ─ ─ ─ ─ ─ ─ ─ ─ ─ ─ ─ ─ ─ ─ ─ ─ ─ ─ ─ ─ ─ ─ ─ ─ ─
architecture top_level of io_controller is

    . . .

begin
    internal_clock_gen : entity work.phase_locked_clock_gen(std_cell)
        port map ( reference => ref_clock,
                    phi1 => work.clock_pkg.clock_phase1,
                    phi2 => work.clock_pkg.clock_phase2 );

    the_bus_sequencer : entity work.bus_sequencer(fsm)
        port map ( rd, wr, sel, width, burst, addr(1 downto 0), ready,
                    control_reg_wr, status_reg_rd, data_fifo_wr, data_fifo_rd,
                    . . . );

    . . .

end architecture top_level;
```

An outline of the entity and architecture body for the input/output controller integrated circuit. The architecture body uses the master clock signals.

FIGURE 7-5

```
architecture fsm of bus_sequencer is
    – – This architecture implements the sequencer as a finite state machine.
    – – NOTE: it uses the clock signals from clock_pkg to synchronize the fsm.
    signal next_state_vector : . . .;
begin
    bus_sequencer_state_register : entity work.state_register(std_cell)
        port map ( phi1 => work.clock_pkg.clock_phase1,
                   phi2 => work.clock_pkg.clock_phase2,
                   next_state => next_state_vector,
                   . . . );

    . . .

end architecture fsm;
```

An outline of the architecture body for the bus sequencer of the input/output controller circuit.

Subprograms in Package Declarations

Another kind of declaration that may be included in a package declaration is a subprogram declaration—either a procedure or a function declaration. This ability allows us to write subprograms that implement useful operations and to call them from a number of different modules. An important use of this feature is to declare subprograms that operate on values of a type declared by the package. This gives us a way of conceptually extending VHDL with new types and operations, so-called *abstract data types*.

An important aspect of declaring a subprogram in a package declaration is that we only write the header of the subprogram, that is, the part that includes the name and the interface list defining the parameters (and result type for functions). We leave out the body of the subprogram. The reason for this is that the package declaration, as we mentioned earlier, provides only the external view of the items it declares, leaving the implementation of the items to the package body. For items such as types and signals, the complete definition is needed in the external view. However, for subprograms, we need only know the information contained in the header to be able to call the subprogram. As users of a subprogram, we need not be concerned with how it achieves its effect or calculates its result. This is an example of a general principle called *information hiding*: making an interface visible but hiding the details of implementation. To illustrate this idea, suppose we have a package declaration that defines a bit-vector subtype:

```
subtype word32 is bit_vector(31 downto 0);
```

We can include in the package a procedure to do addition on **word32** values that represent signed integers. The procedure declaration in the package declaration is

```
procedure add ( a, b : in word32;
               result : out word32;  overflow : out boolean );
```

Note that we do not include the keyword **is** or any of the local declarations or statements needed to perform the addition. These are deferred to the package body. All we include is the description of the formal parameters of the procedure. Similarly, we might include a function to perform an arithmetic comparison of two **word32** values:

function ”<” (a, b : **in** word32) **return** boolean;

Again, we omit the local declarations and statements, simply specifying the formal parameters and the result type of the function.

Constants in Package Declarations

Just as we can apply the principle of information hiding to subprograms declared in a package, we can also apply it to constants declared in a package. The external view of a constant is just its name and type. We need to know these in order to use it, but we do not actually need to know its value. This may seem strange at first, but if we recall that the idea of introducing constant declarations in the first place was to avoid scattering literal values throughout a model, it makes more sense. We defer specifying the value of a constant declared in a package by omitting the initialization expression, for example:

constant max_buffer_size : positive;

This defines the constant to be a positive integer value. However, since we cannot see the actual value, we are not tempted to write the value as an integer literal in a model that uses the package. The specification of the actual value is deferred to the package body, where it is not visible to a model that uses the package. Given the above deferred constant in a package declaration, the corresponding package body must include the full constant declaration, for example:

constant max_buffer_size : positive := 4096;

Note that we do not have to defer the value in a constant declaration—it is optional.

EXAMPLE

We can extend the package specification from Figure 7-1, declaring useful types for a CPU model, by including declarations related to opcode processing. The revised package is shown in Figure 7-6. It includes a subtype that represents an opcode value, a function to extract an opcode from an instruction word and a number of constants representing the opcodes for different instructions.

Figure 7-7 shows a behavioral model of a CPU that uses these declarations. The instruction set interpreter process declares a variable of the **opcode** type and uses the **extract_opcode** function to extract the bits representing the opcode from the fetched instruction word. It then uses the constants from the package as choices in a case statement to decode and execute the instruction specified by the opcode.

FIGURE 7-6

```
package cpu_types is

    constant word_size : positive := 16;
    constant address_size : positive := 24;

    subtype word is bit_vector(word_size – 1 downto 0);
    subtype address is bit_vector(address_size – 1 downto 0);

    type status_value is ( halted, idle, fetch, mem_read, mem_write,
                           io_read, io_write, int_ack );

    subtype opcode is bit_vector(5 downto 0);

    function extract_opcode ( instr_word : word ) return opcode;

    constant op_nop : opcode := "000000";
    constant op_breq : opcode := "000001";
    constant op_brne : opcode := "000010";
    constant  op_add : opcode := "000011";   . . .

end package cpu_types;
```

A revised version of the package used in a CPU model.

FIGURE 7-7

```
architecture behavioral of cpu is
begin
    interpreter : process is
        variable instr_reg : work.cpu_types.word;
        variable instr_opcode : work.cpu_types.opcode;
    begin
        . . .        – – initialize
        loop
            . . .        – – fetch instruction
            instr_opcode := work.cpu_types.extract_opcode ( instr_reg );
            case instr_opcode is
                when work.cpu_types.op_nop => null;
                when work.cpu_types.op_breq => . . .
                    . . .
            end case;
        end loop;
    end process interpreter;
end architecture behavioral;
```

An outline of a CPU model that uses items declared in the revised cpu_types *package.*

Note that since the constants are used as choices in the case statement, they must be locally static. If we had deferred the values of the constants to the package body, their value would not be known when the case statement was analyzed. This is why we included the constant values in the package declaration. In general, the value of a deferred constant is not locally static.

7.2 Package Bodies

Now that we have seen how to define the interface to a package, we can turn to the package body. Each package declaration that includes subprogram declarations or deferred constant declarations must have a corresponding package body to fill in the missing details. However, if a package declaration only includes other kinds of declarations, such as types, signals or fully specified constants, no package body is necessary. The syntax rule for a package body is similar to that for the interface, but with the inclusion of the keyword **body**:

> package_body ⇐
>   **package body** identifier **is**
>   { package_body_declarative_item }
>   **end** [**package body**] [identifier] ;

The items declared in a package body must include the full declarations of all subprograms defined in the corresponding package declaration. These full declarations must include the subprogram headers exactly as they are written in the package declaration, to ensure that the implementation *conforms* with the interface. This means that the names, types, modes and default values of each of the formal parameters must be repeated exactly. There are only two variations allowed. First, a numeric literal may be written differently, for example, in a different base, provided it has the same value. Second, a simple name consisting just of an identifier may be replaced by a selected name, provided it refers to the same item. While this conformance requirement might seem an imposition at first, in practice it is not. Any reasonable text editor used to create a VHDL model allows the header to be copied from the package declaration with little difficulty. Similarly, a deferred constant defined in a package declaration must have its value specified by repeating the declaration in the package body, this time filling in the initialization expression as in a full constant declaration.

In addition to the full declarations of items deferred from the package declaration, a package body may include declarations of additional types, subtypes, constants and subprograms. These items are used to implement the subprograms defined in the package declaration. Note that the items declared in the package declaration cannot be declared again in the body (apart from subprograms and deferred constants, as described above), since they are automatically visible in the body. Furthermore, the package body cannot include declarations of additional signals. Signal declarations may only be included in the interface declaration of a package.

EXAMPLE

Figure 7-8 shows outlines of a package declaration and a package body declaring overloaded versions of arithmetic operators for bit-vector values. The functions treat bit vectors as representing signed integers in binary form. Only the function headers are included in the package declaration. The package body contains the full function bodies. It also includes a function, mult_unsigned, not defined in the package declaration. It is used internally in the package body to implement the signed multiplication operator.

FIGURE 7-8

```
package bit_vector_signed_arithmetic is
    function "+" ( bv1, bv2 : bit_vector ) return bit_vector;
    function "–" ( bv : bit_vector ) return bit_vector;
    function "*" ( bv1, bv2 : bit_vector ) return bit_vector;
    . . .
end package bit_vector_signed_arithmetic;
_ _ _ _ _ _ _ _ _ _ _ _ _ _ _ _ _ _ _ _ _ _ _ _ _ _ _ _ _ _ _ _ _ _ _ _ _ _ _
package body bit_vector_signed_arithmetic is
    function "+" ( bv1, bv2 : bit_vector ) return bit_vector is . . .
    function "–" ( bv : bit_vector ) return bit_vector is . . .
    function mult_unsigned ( bv1, bv2 : bit_vector ) return bit_vector is
        . . .
    begin
        . . .
    end function mult_unsigned;
    function "*" ( bv1, bv2 : bit_vector ) return bit_vector is
    begin
        if bv1(bv1'left) = '0' and bv2(bv2'left) = '0' then
            return mult_unsigned(bv1, bv2);
        elsif bv1(bv1'left) = '0' and bv2(bv2'left) = '1' then
            return –mult_unsigned(bv1, –bv2);
        elsif bv1(bv1'left) = '1' and bv2(bv2'left) = '0' then
            return –mult_unsigned(–bv1, bv2);
        else
            return mult_unsigned(–bv1, –bv2);
        end if;
    end function "*";

    . . .
end package body bit_vector_signed_arithmetic;
```

An outline of a package declaration and body that define signed arithmetic functions on integers represented as bit vectors.

VHDL-87

The keywords **package body** may not be included at the end of a package body in VHDL-87.

One final point to mention on the topic of packages relates to the order of analysis. We mentioned before that a package is a separate design unit that is analyzed separately from other design units, such as entity declarations and architecture bodies. In fact, a package declaration and its corresponding package body are each separate design units, hence they may be analyzed separately. A package declaration is a primary design unit, and a package body is a secondary design unit. The package body depends on information defined in the package declaration, so the declaration must be analyzed first. Furthermore, the declaration must be analyzed before any other design unit that refers to an item defined by the package. Once the declaration has been analyzed, it does not matter when the body is analyzed in relation to units that use the package, provided it is analyzed before the model is elaborated. In a large suite of models, the dependency relationships can get quite complex, and a correct order of analysis can be difficult to find. A good VHDL tool suite will provide some degree of automating this process by working out the dependency relationships and analyzing those units needed to build a particular target unit to simulate or synthesize.

7.3   Use Clauses

We have seen how we can refer to an item provided by a package by writing its selected name, for example, work.cpu_types.status_value. This name refers to the item status_value in the package cpu_types stored in the library work. If we need to refer to this object in many places in a model, having to write the library name and package name becomes tedious and can obscure the intent of the model. We saw in Chapter 5 that we can write a *use clause* to make a library unit directly visible in a model, allowing us to omit the library name when referring to the library unit. Since an analyzed package is a library unit, use clauses also apply to making packages directly visible. So we could precede a model with a use clause referring to the package defined in the example in Figure 7-1 on page 196:

> **use** work.cpu_types;

This use clause allows us to write declarations in our model more simply, for example:

> **variable** data_word : cpu_types.word;
> **variable** next_address : cpu_types.address;

In fact, the use clause is more general than this usage indicates and allows us to make any name from a library or package directly visible. Let us look at the full syntax rule for a use clause, then discuss some of the possibilities.

> use_clause ⇐ **use** selected_name { , ... } ;
>
> selected_name ⇐
>   name . (identifier ‖ character_literal ‖ operator_symbol ‖ **all**)

The syntax rule for names, shown in Appendix C, includes the possibility of a name itself being either a selected name or a simple identifier. If we make these substitutions in the above syntax rule, we see that a selected name can be of the form

> identifier . identifier . (identifier ‖ character_literal ‖ operator_symbol ‖ **all**)

One possibility is that the first identifier is a library name, and the second is the name of a package within the library. This form allows us to refer directly to items within a package without having to use the full selected name. For example, we can simplify the above declarations even further by rewriting the use clause as

> **use** work.cpu_types.word, work.cpu_types.address;

The declarations can then be written as

```
variable data_word : word;
variable next_address : address;
```

We can place a use clause in any declarative part in a model. One way to think of a use clause is that it "imports" the names of the listed items into the part of the model containing the use clause, so that they can be used without writing the library or package name. The names become directly visible after the use clause, according to the same visibility rules that we discussed in Chapter 6.

The syntax rule for a use clause shows that we can write the keyword **all** instead of the name of a particular item to import from a package. This form is very useful, as it is a shorthand way of importing all of the names defined in the interface of a package. For example, if we are using the IEEE standard-logic package as the basis for the data types in a design, it is often convenient to import everything from the standard-logic package, including all of the overloaded operator definitions. We can do this with a use clause as follows:

```
use ieee.std_logic_1164.all;
```

This use clause means that the model imports all of the names defined in the package std_logic_1164 residing in the library **ieee**. This explains the "magic" that we have used in previous chapters when we needed to model data using the standard-logic types. The keyword **all** can be included for any package where we want to import all of the declarations from the package into a model.

EXAMPLE

Figure 7-9 is a revised version of the architecture body outlined in Figure 7-7 on page 202. It includes a use clause referring to items declared in the **cpu_types** package. This makes the rest of the model considerably less cluttered and easier to read. The use clause is included within the declarative part of the instruction set interpreter process. Thus the names "imported" from the package are directly visible in the rest of the declarative part and in the body of the process.

FIGURE 7-9

```
architecture behavioral of cpu is
begin
    interpreter : process is
        use work.cpu_types.all;
        variable instr_reg : word;
        variable instr_opcode : opcode;
    begin
        ...        -- initialize
        loop
            ...        -- fetch instruction
            instr_opcode := extract_opcode ( instr_reg );
            case instr_opcode is
                when op_nop => null;
```

```
                        when op_breq => . . .
                              . . .
                     end case;
                  end loop;
              end process interpreter;
       end architecture behavioral;
```

A revised outline of a CPU model, including a use clause to refer to items from the cpu_types package.

One final point to clarify about use clauses before looking at an extended example is the way in which they may be included at the beginning of a design unit, as well as in declarative parts within a library unit. We have seen in Section 5.5 how we may include library and use clauses at the head of a design unit, such as an entity interface or architecture body. This area of a design unit is called its *context clause*. In fact, this is probably the most common place for including use clauses. The names imported here are made directly visible throughout the design unit. For example, if we want to use the IEEE standard-logic type std_ulogic in the declaration of an entity, we might write the design unit as follows:

```
library ieee;  use ieee.std_logic_1164.std_ulogic;

entity logic_block is
     port ( a, b : in std_ulogic;
               y, z : out std_ulogic );
end entity logic_block;
```

The library clause and the use clause together form the context clause for the entity declaration in this example. The library clause makes the contents of the library accessible to the model, and the use clause imports the type name std_ulogic declared in the package std_logic_1164 in the library ieee. By including the use clause in the context clause of the entity declaration, the std_ulogic type name is available when declaring the ports of the entity.

The names imported by a use clause in this way are made directly visible in the entire design unit after the use clause. In addition, if the design unit is a primary unit (such as an entity declaration or a package declaration), the visibility is extended to any corresponding secondary unit. Thus, if we include a use clause in the primary unit, we do not need to repeat it in the secondary unit, as the names are automatically visible there.

7.4   The Predefined Package Standard

In previous chapters, we have introduced numerous predefined types and operators. We can use them in our VHDL models without having to write type declarations or subprogram definitions for them. These predefined items all come from a special package called standard, located in a special design library called std. A full listing of the standard package is included for reference in Appendix A.

Because nearly every model we write needs to make use of the contents of this library and package, as well as the library **work**, VHDL includes an implicit context clause of the form

library std, work; **use** std.standard.**all**;

at the beginning of each design unit. Hence we can refer to the simple names of the predefined items without having to resort to their selected names. In the occasional case where we need to distinguish a reference to a predefined operator from an overloaded version, we can use a selected name, for example:

result := std.standard."<" (a, b);

EXAMPLE

A package that provides signed arithmetic operations on integers represented as bit vectors might include a relational operator, defined as shown in Figure 7-10. The function negates the sign bit of each operand, then compares the resultant bit vectors using the predefined relational operator from the package **standard**. The full selected name for the predefined operator is necessary to distinguish it from the function being defined. If the return expression were written as "**tmp < tmp2**", it would refer to the function in which it occurs, creating a circular definition.

FIGURE 7-10

```
function "<" ( a, b : bit_vector ) return boolean is
    variable tmp1 : bit_vector(a'range) := a;
    variable tmp2 : bit_vector(b'range) := b;
begin
    tmp1(tmp1'left) := not tmp1(tmp1'left);
    tmp2(tmp2'left) := not tmp2(tmp2'left);
    return std.standard."<" ( tmp1, tmp2 );
end function "<";
```

An operator function for comparing two bit vectors representing signed integers.

Exercises

1. [❶ 7.1] Write a package declaration for use in a model of an engine management system. The package contains declarations of a physical type, **engine_speed**, expressed in units of revolutions per minute (RPM); a constant, **peak_rpm**, with a value of 6000 RPM; and an enumeration type, **gear**, with values representing first, second, third, fourth and reverse gears. Assuming the package is analyzed and stored in the current working library, write selected names for each of the items declared in the package.

2. [❶ 7.1] Write a declaration for a procedure that increments an integer, as the procedure declaration would appear in a package declaration.

3. [❶ 7.1] Write a declaration for a function that tests whether an integer is odd, as the function declaration would appear in a package declaration.

4. [❶ 7.1] Write a deferred constant declaration for the real constant *e* = 2.71828.

5. [❶ 7.2] Is a package body required for the package declaration described in Exercise 1?

6. [❶ 7.3] Write a use clause that makes the engine_speed type from the package described in Exercise 1 directly visible.

7. [❶ 7.3] Write a context clause that makes a library DSP_lib accessible and that makes an entity systolic_FFT and all items declared in a package DSP_types in the library directly visible.

8. [❷ 7.4] Integers can be represented in *signed magnitude* form, in which the leftmost bit represents the sign ('0' for non-negative, '1' for negative), and the remaining bits are the absolute value of the number, represented in binary. If we wish to compare bit vectors containing numbers in signed magnitude form, we cannot use the predefined relational operators directly. We must first transform each number as follows: if the number is negative, complement all bits; if the number is non-negative, complement only the sign bit. Write a comparison function, overloading the operator "<", to compare signed-magnitude bit vectors using this method.

9. [❸ 7.1/7.2] Develop a package declaration and body that provide operations for dealing with time-of-day values. The package defines a time-of-day value as a record containing hours, minutes and seconds since midnight and provides deferred constants representing midnight and midday. The operations provided by the package are

   • comparison ("<", ">", "<=" and ">="),

   • addition of a time-of-day value and a number of seconds to yield a time-of-day result, and

   • subtraction of two time-of-day values to yield a number-of-seconds result.

10. [❸ 7.1/7.2] Develop a package declaration and body to provide operations on character strings representing identifiers. An outline of the package declaration is

```
package identifier_pkg is

    subtype identifier is string(1 to 15);

    constant max_table_size : integer := 50;
    subtype table_index is integer range 1 to max_table_size;
    type table is array (table_index) of identifier;

    . . .

end package identifier_pkg;
```

The package also declares a procedure to convert alphabetic characters in a string to lowercase and a procedure to search for an occurrence of a given identifier in a table. The search procedure has two **out** mode parameters: a Boolean value indicating whether the sought string is in the table and a table_index value indicating its position, if present.

Resolved Signals

Throughout the previous chapters we have studiously avoided considering the case of multiple output ports connecting the one signal. The problem that arises in such a case is determining the final value of the signal when multiple sources drive it. In this chapter we discuss *resolved signals*, the mechanism provided by VHDL for modeling such cases.

8.1   Basic Resolved Signals

If we consider a real digital system with two outputs driving one signal, we can fairly readily determine the resulting value based on some analog circuit theory. The signal is driven to some intermediate state, depending on the drive capacities of the conflicting drivers. This intermediate state may or may not represent a valid logic state. Usually we only connect outputs in a design if at most one is active at a time, and the rest are in some high-impedance state. In this case, the resulting value should be the driving value of the single active output. In addition, we include some form of "pull-up" that determines the value of the signal when all outputs are inactive.

While this simple approach is satisfactory for some models, there are other cases where we need to go further. One of the reasons for simulating a model of a design is to detect errors such as multiple simultaneously active connected outputs. In this case, we need to extend the simple approach to detect such errors. Another problem arises when we are modeling at a higher level of abstraction and are using more complex types. We need to specify what, if anything, it means to connect multiple outputs of an enumeration type together.

The approach taken by VHDL is a very general one: the language requires the designer to specify precisely what value results from connecting multiple outputs. It does this through *resolved signals*, which are an extension of the basic signals we have used in previous chapters. A resolved signal includes in its definition a function, called the *resolution function*, that is used to calculate the final signal value from the values of all of its sources.

Let us see how this works by developing an example. We can model the values driven by a tristate output using a simple extension to the predefined type **bit**, for example:

type tri_state_logic **is** ('0', '1', 'Z');

The extra value, 'Z', is used by an output to indicate that it is in the high-impedance state. Next, we need to write a function that takes a collection of values of this type, representing the values driven by a number of outputs, and return the resulting value to be applied to the connected signal. For this example, we assume that at most one driver is active ('0' or '1') at a time and that the rest are all driving 'Z'. The difficulty with writing the function is that we should not restrict it to a fixed number of input values. We can avoid this by giving it a single parameter that is an unconstrained array of tri_state_logic values, defined by the type declaration

type tri_state_logic_array **is array** (integer **range** <>) **of** tri_state_logic;

The declaration of the resolution function is shown in Figure 8-1. The final step to making a resolved signal is to declare the signal, as follows:

signal s1 : resolve_tri_state_logic tri_state_logic;

This declaration is almost identical to a normal signal declaration, but with the addition of the resolution function name before the signal type. The signal still takes on values from the type **tri_state_logic**, but inclusion of a function name indicates that the signal is a resolved signal, with the named function acting as the resolution function. The fact that **s1** is resolved means that we are allowed to have more than one

FIGURE 8-1

```
function resolve_tri_state_logic ( values : in tri_state_logic_array )
                                return tri_state_logic is
    variable result : tri_state_logic := 'Z';
begin
    for index in values'range loop
        if values(index) /= 'Z' then
            result := values(index);
        end if;
    end loop;
    return result;
end function resolve_tri_state_logic;
```

A resolution for resolving multiple values from tristate drivers.

source for it in the design. (Sources include drivers within processes and output ports of components associated with the signal.) When a transaction is scheduled for the signal, the value is not applied to the signal directly. Instead, the values of all sources connected to the signal, including the new value from the transaction, are formed into an array and passed to the resolution function. The result returned by the function is then applied to the signal as its new value.

Let us look at the syntax rule that describes the VHDL mechanism we have used in the above example. It is an extension of the rules for the subtype indication, which we first introduced in Chapters 2 and 4. The combined rule is

```
subtype_indication ⇐
    [ resolution_function_name ]
    type_mark
    [ range ( range_attribute_name
            ‖ simple_expression ( to ‖ downto ) simple_expression )
    ‖ ( discrete_range { , ... } ) ]
```

This rule shows that a subtype indication can optionally include the name of a function to be used as a resolution function. Given this new rule, we can include a resolution function name anywhere that we specify a type to be used for a signal. For example, we could write a separate subtype declaration that includes a resolution function name, defining a *resolved subtype*, then use this subtype to declare a number of resolved signals, as follows:

subtype resolved_logic **is** resolve_tri_state_logic tri_state_logic;

signal s2, s3 : resolved_logic;

The subtype **resolved_logic** is a resolved subtype of tri_state_logic, with resolve_tri_state_logic acting as the resolution function. The signals s2 and s3 are resolved signals of this subtype. Where a design makes extensive use of resolved signals, it is good practice to define resolved subtypes and use them to declare the signals and ports in the design.

The resolution function for a resolved signal is also invoked to initialize the signal. At the start of a simulation, the drivers for the signal are initialized to the expression included in the signal declaration, or to the default initial value for the signal type if no initialization expression is given. The resolution function is then invoked using these driver values to determine the initial value for the signal. In this way, the signal always has a properly resolved value, right from the start of the simulation.

Let us now return to the tristate logic type we introduced earlier. In the previous example, we assumed that at most one driver is '0' or '1' at a time. In a more realistic model, we need to deal with the possibility of driver conflicts, in which one source drives a resolved signal with the value '0' and another drives it with the value '1'. In some logic families, such driver conflicts cause an indeterminate signal value. We can represent this indeterminate state with a fourth value of the logic type, 'X', often called an *unknown* value. This gives us a complete and consistent *multivalued logic* type, which we can use to describe signal values in a design in more detail than we can using just bit values.

EXAMPLE

Figure 8-2 shows a package interface and the corresponding package body for the four-state multivalued logic type. The constant resolution_table is a lookup table used to determine the value resulting from two source contributions to a signal of the resolved logic type. The resolution function uses this table, indexing it with each element of the array passed to the function. If any source contributes 'X', or if there are two sources with conflicting '0' and '1' contributions, the result is 'X'. If one or more sources are '0' and the remainder 'Z', the result is '0'. Similarly, if one or more sources are '1' and the remainder 'Z', the result is '1'. If all sources are 'Z', the result is 'Z'. The lookup table is a compact way of representing this set of rules.

FIGURE 8-2

```
package MVL4 is
    type MVL4_ulogic is ('X', '0', '1', 'Z');  -- unresolved logic type
    type MVL4_ulogic_vector is array (natural range <>) of MVL4_ulogic;
    function resolve_MVL4 ( contribution : MVL4_ulogic_vector )
                            return MVL4_ulogic;
    subtype MVL4_logic is resolve_MVL4 MVL4_ulogic;
end package MVL4;

--------------------------------------------------

package body MVL4 is
    type table is array (MVL4_ulogic, MVL4_ulogic) of MVL4_ulogic;
    constant resolution_table : table :=
        --  'X'  '0'  '1'  'Z'
        --  ----------------
        ( ( 'X', 'X', 'X', 'X' ),   -- 'X'
          ( 'X', '0', 'X', '0' ),   -- '0'
          ( 'X', 'X', '1', '1' ),   -- '1'
          ( 'X', '0', '1', 'Z' ) ); -- 'Z'
```

```
      function resolve_MVL4 ( contribution : MVL4_ulogic_vector )
                         return MVL4_ulogic is
         variable result : MVL4_ulogic := 'Z';
      begin
         for index in contribution'range loop
            result := resolution_table(result, contribution(index));
         end loop;
         return result;
      end function resolve_MVL4;
   end package body MVL4;
```

A package interface and body for a four-state multivalued and resolved logic subtype.

We can use this package in a design for a tristate buffer. The entity declaration and a behavioral architecture body are shown in Figure 8-3. The buffer drives the value 'Z' on its output when it is disabled. It copies the input to the output when it is enabled and the input is a proper logic level ('0' or '1'). If either the input or the enable port is not a proper logic level, the buffer drives the unknown value on its output.

FIGURE 8-3

```
use work.MVL4.all;
entity tri_state_buffer is
   port ( a, enable : in MVL4_ulogic;  y : out MVL4_ulogic );
end entity tri_state_buffer;
- - - - - - - - - - - - - - - - - - - - - - - - - - - - - - - - - - - - - - -
architecture behavioral of tri_state_buffer is
begin
   y <= 'Z' when enable = '0' else
        a   when enable = '1' and (a = '0' or a = '1') else
        'X';
end architecture behavioral;
```

An entity and behavioral architecture body for a tristate buffer.

Figure 8-4 shows the outline of an architecture body that uses the tristate buffer. The signal **selected_val** is a resolved signal of the multivalued logic type. It is driven by the two buffer output ports. The resolution function for the signal is used to determine the final value of the signal whenever a new transaction is applied to either of the buffer outputs.

FIGURE 8-4

```
use work.MVL4.all;

architecture gate_level of misc_logic is
    signal src1, src1_enable : MVL4_ulogic;
    signal src2, src2_enable : MVL4_ulogic;
    signal selected_val : MVL4_logic;
    . . .

begin
    src1_buffer : entity work.tri_state_buffer(behavioral)
        port map ( a => src1, enable => src1_enable, y => selected_val );
    src2_buffer : entity work.tri_state_buffer(behavioral)
        port map ( a => src2, enable => src2_enable, y => selected_val );
    . . .

end architecture gate_level;
```

An outline of an architecture body that uses the tristate buffer. The output ports of the two instances of the buffer form two sources for the resolved signal selected_val.

Composite Resolved Subtypes

The above examples have all shown resolved subtypes of scalar enumeration types. In fact, VHDL's resolution mechanism is more general. We can use it to define a resolved subtype of any type that we can legally use as the type of a signal. Thus, we can define resolved integer subtypes, resolved composite subtypes and others. In the latter case, the resolution function is passed an array of composite values and must determine the final composite value to be applied to the signal.

EXAMPLE

Figure 8-5 shows a package interface and body that define a resolved array subtype. Each element of an array value of this subtype can be 'X', '0', '1' or 'Z'. The unresolved type uword is a 32-element array of these values. The resolution function has an unconstrained array parameter consisting of elements of type uword. The function uses the lookup table to resolve corresponding elements from each of the contributing sources and produces a 32-element array result. The subtype word is the final resolved array subtype.

FIGURE 8-5

```
package words is
    type X01Z is ('X', '0', '1', 'Z');
    type uword is array (0 to 31) of X01Z;
    type uword_vector is array (natural range <>) of uword;
    function resolve_word ( contribution : uword_vector ) return uword;
    subtype word is resolve_word uword;
end package words;
```

```
- - - - - - - - - - - - - - - - - - - - - - - - - - - - - - - - - - - - - - - -
package body words is
    type table is array (X01Z, X01Z) of X01Z;
    constant resolution_table : table :=
        -- 'X'  '0'  '1'  'Z'
        -- - - - - - - - - - - -
        ( (  'X',  'X',  'X',  'X'  ),     -- 'X'
          (  'X',  '0',  'X',  '0'  ),     -- '0'
          (  'X',  'X',  '1',  '1'  ),     -- '1'
          (  'X',  '0',  '1',  'Z'  ) );   -- 'Z'
    function resolve_word ( contribution : uword_vector ) return uword is
        variable result : uword := (others => 'Z');
    begin
        for index in contribution'range loop
            for element in uword'range loop
                result(element) :=
                        resolution_table( result(element), contribution(index)(element) );
            end loop;
        end loop;
        return result;
    end function resolve_word;
end package body words;
```

A package interface and body for a resolved array subtype.

We can use these types to declare array ports in entity declarations and resolved array signals with multiple sources. Figure 8-6 shows outlines of a CPU entity and a memory entity, which have bidirectional data ports of the unresolved array type. The architecture body for a computer system, also outlined in Figure 8-6, declares a signal of the resolved subtype and connects it to the data ports of the instances of the CPU and memory.

FIGURE 8-6 _____

```
use work.words.all;
entity cpu is
    port ( address : out uword;  data : inout uword;  . . . );
end entity cpu;

- - - - - - - - - - - - - - - - - - - - - - - - - - - - - - - - - - - - - - - -

use work.words.all;
entity memory is
    port ( address : in uword;  data : inout uword; . . . );
end entity memory;

- - - - - - - - - - - - - - - - - - - - - - - - - - - - - - - - - - - - - - - -
```

(continued on page 218)

(continued from page 217)

```
architecture top_level of computer_system is
    use work.words.all;
    signal address : uword;
    signal data : word;
    . . .
begin
    the_cpu : entity work.cpu(behavioral)
        port map ( address, data, . . . );
    the_memory : entity work.memory(behavioral)
        port map ( address, data, . . . );

    . . .

end architecture top_level;
```

An outline of a CPU and memory entity with resolved array ports, and an architecture body for a computer system that uses the CPU and memory.

A resolved composite subtype works well provided every source for a resolved signal of the subtype is connected to every element of the signal. For the subtype shown in the example, every source must be a 32-element array and must connect to all 32 elements of the data signal. However, in a realistic computer system, sources are not always connected in this way. For example, we may wish to connect an eight-bit-wide device to the low-order eight bits of a 32-bit-wide data bus. We might attempt to express such a connection in a component instantiation statement, as follows:

```
boot_rom : entity work.ROM(behavioral)
    port map ( a => address, d => data(24 to 31), . . . ); – – illegal
```

If we add this statement to the architecture body in Figure 8-6, we have two sources for elements 0 to 23 of the data signal and three for elements 24 to 31. A problem arises when resolving the signal, since we are unable to construct an array containing the contributions from the sources. For this reason, VHDL does not allow us to write such a description; it is illegal.

The solution to this problem is to describe the data signal as an array of resolved elements, rather than as a resolved array of elements. We can declare an array type whose elements are values of the MVL4_logic type, shown in Figure 8-2. The array type declaration is

```
type MVL4_logic_vector is array (natural range <>) of MVL4_logic;
```

This approach has the added advantage that the array type is unconstrained, so we can use it to create signals of different widths, each element of which is resolved. An important point to note, however, is that the type MVL4_logic_vector is distinct from the type MVL4_ulogic_vector, since they are defined by separate type declarations. Neither is a subtype of the other. Hence we cannot legally associate a signal of type MVL4_logic_vector with a port of type MVL4_ulogic_vector, or a signal of type MVL4_ulogic_vector with a port of type MVL4_logic_vector. One solution is to identify all ports that may need to be associated with a signal of the resolved type and to declare them to be of the resolved type. This avoids the type mismatch that would otherwise occur. We illustrate this approach in the following example.

EXAMPLE

Let us assume that the type **MVL4_logic_vector** described above has been added to the package **MVL4**. Figure 8-7 shows entity declarations for a ROM entity and a single in-line memory module (SIMM), using the **MVL4_logic_vector** type for their data ports. The data port of the SIMM is 32 bits wide, whereas the data port of the ROM is parameterized by a generic constant.

FIGURE 8-7

```
use work.MVL4.all;
entity ROM is
    port ( a : in MVL4_ulogic_vector(15 downto 0);
           d : inout MVL4_logic_vector(7 downto 0);
           rd : in MVL4_ulogic );
end entity ROM;

-------------------------------------------------

use work.MVL4.all;
entity SIMM is
    port ( a : in MVL4_ulogic_vector(9 downto 0);
           d : inout MVL4_logic_vector(31 downto 0);
           ras, cas, we, cs : in MVL4_ulogic );
end entity SIMM;

-------------------------------------------------

architecture detailed of memory_subsystem is
    signal internal_data : MVL4_logic_vector(31 downto 0);
    . . .
begin
    boot_ROM : entity work.ROM(behavioral)
        port map ( a => internal_addr(15 downto 0),
                   d => internal_data(7 downto 0),
                   rd => ROM_select );

    main_mem : entity work.SIMM(behavioral)
        port map ( a => main_mem_addr, d => internal_data, . . . );

    . . .

end architecture detailed;
```

Entity declarations for memory modules whose data ports are arrays of resolved elements, and an outline of an architecture body that uses these entities.

Figure 8-7 also shows an outline of an architecture body that uses these two entities. It declares a signal, internal_data, of the MVL4_logic_vector type, representing 32 individually resolved elements. The SIMM entity is instantiated with its data port connected to all 32 internal data elements. The ROM entity is instantiated with the d_width generic constant set to eight, which constrains the data port of the instance to eight elements. These are connected to the rightmost eight elements of the internal data signal. When any of these elements is resolved, the resolution function is passed contributions from the corresponding elements of the SIMM and ROM data ports. When any of the remaining elements of the internal data signal are resolved, they have one less contribution, since they are not connected to any element of the ROM data port.

Summary of Resolved Subtypes

At this point, let us summarize the important points about resolved signals and their resolution functions. Resolved signals of resolved subtypes are the only means by which we may connect a number of sources together, since we need a resolution function to determine the final value of the signal or port from the contributing values. The resolution function must take a single parameter that is a one-dimensional unconstrained array of values of the signal type, and must return a value of the signal type. The index type of the array does not matter, so long as it contains enough index values for the largest possible collection of sources connected together. For example, an array type declared as follows is inadequate if the resolved signal has five sources:

```
type small_int is range 1 to 4;
type small_array is array (small_int range <>) of . . . ;
```

The resolution function must be a pure function; that is, it must not have any side effects. This requirement is a safety measure to ensure that the function always returns a predictable value for a given set of source values. Furthermore, since the source values may be passed in any order within the array, the function should be commutative; that is, its result should be independent of the order of the values. When the design is simulated, the resolution function is called whenever any of the resolved signal's sources is active. The function is passed an array of all of the current source values and the result it returns is used to update the signal value. When the design is synthesized, the resolution function specifies the way in which the synthesized hardware should combine values from multiple sources for a resolved signal.

8.2 IEEE Std_Logic_1164 Resolved Subtypes

In previous chapters we have used the IEEE standard multivalued logic package, std_logic_1164. We are now in a position to describe all of the items provided by the package, including the resolved subtypes and operators. The intent of the IEEE standard is that the multivalued logic subtypes defined in the package be used for models that must be interchanged between designers. The full package interface is included for reference in Appendix B. First, recall that the package provides the basic type std_ulogic, defined as

type std_ulogic **is** ('U', 'X', '0', '1', 'Z', 'W', 'L', 'H', '–');

and an array type **std_ulogic_vector**, defined as

type std_ulogic_vector **is array** (natural **range** <>) **of** std_ulogic;

We have not mentioned it before, but the "u" in "ulogic" stands for unresolved. These types serve as the basis for the declaration of the resolved subtype **std_logic**, defined as follows:

function resolved (s : std_ulogic_vector) **return** std_ulogic;

subtype std_logic **is** resolved std_ulogic;

The standard-logic package also declares an array type of standard-logic elements, analogous to the **bit_vector** type, for use in declaring array signals:

type std_logic_vector **is array** (natural **range** <>) **of** std_logic;

The IEEE standard recommends that models use the subtype **std_logic** and the type **std_logic_vector** instead of the unresolved types **std_ulogic** and **std_ulogic_vector**, even if a signal has only one source. The reason is that simulation vendors are expected to optimize simulation of models using the resolved subtype, but need not optimize use of the unresolved type. The disadvantage of this approach is that it prevents detection of erroneous designs in which multiple sources are inadvertently connected to a signal that should have only one source. Nevertheless, if we are to conform to the standard practice, we should use the resolved logic type. We will conform to the standard in the subsequent examples in this book.

The standard defines the resolution function **resolved** as shown in Figure 8-8. VHDL tools are allowed to provide built-in implementations of this function to improve performance. The function uses the constant **resolution_table** to resolve the driving values. If there is only one driving value, the function returns that value unchanged. If the function is passed an empty array, it returns the value 'Z'. The value of **resolution_table** shows exactly what is meant by "forcing" driving values ('X', '0' and '1') and "weak" driving values ('W', 'L' and 'H'). If one driver of a resolved signal drives a forcing value and another drives a weak value, the forcing value dominates. On the other hand, if both drivers drive different values with the same strength, the result is the unknown value of that strength ('X' or 'W'). The high-impedance value, 'Z', is dominated by forcing and weak values. If a "don't care" value ('–') is to be resolved with any other value, the result is the unknown value 'X'. The interpretation of the "don't care" value is that the model has not made a choice about its output state. Finally, if an "uninitialized" value ('U') is to be resolved with any other value, the result is 'U', indicating that the model has not properly initialized all outputs.

In addition to this multivalued logic subtype, the package **std_logic_1164** declares a number of subtypes for more restricted multivalued logic modeling. The subtype declarations are

subtype X01 **is** resolved std_ulogic **range** 'X' **to** '1';   *– – ('X','0','1')*
subtype X01Z **is** resolved std_ulogic **range** 'X' **to** 'Z';   *– – ('X','0','1','Z')*
subtype UX01 **is** resolved std_ulogic **range** 'U' **to** '1';   *– – ('U','X','0','1')*
subtype UX01Z **is** resolved std_ulogic **range** 'U' **to** 'Z'; *– – ('U','X','0','1','Z')*

FIGURE 8-8

```
type stdlogic_table is array (std_ulogic, std_ulogic) of std_ulogic;
constant resolution_table : stdlogic_table :=
    --  -----------------------------
    --  'U', 'X', '0', '1', 'Z', 'W', 'L', 'H', '-'
    --  -----------------------------
    ( ( 'U', 'U', 'U', 'U', 'U', 'U', 'U', 'U', 'U' ),    -- 'U'
      ( 'U', 'X', 'X', 'X', 'X', 'X', 'X', 'X', 'X' ),    -- 'X'
      ( 'U', 'X', '0', 'X', '0', '0', '0', '0', 'X' ),    -- '0'
      ( 'U', 'X', 'X', '1', '1', '1', '1', '1', 'X' ),    -- '1'
      ( 'U', 'X', '0', '1', 'Z', 'W', 'L', 'H', 'X' ),    -- 'Z'
      ( 'U', 'X', '0', '1', 'W', 'W', 'W', 'W', 'X' ),    -- 'W'
      ( 'U', 'X', '0', '1', 'L', 'W', 'L', 'W', 'X' ),    -- 'L'
      ( 'U', 'X', '0', '1', 'H', 'W', 'W', 'H', 'X' ),    -- 'H'
      ( 'U', 'X', 'X', 'X', 'X', 'X', 'X', 'X', 'X' )     -- '-'
    );
function resolved ( s : std_ulogic_vector ) return std_ulogic is
    variable result : std_ulogic := 'Z';  -- weakest state default
begin
    if s'length = 1 then
        return s(s'low);
    else
        for i in s'range loop
            result := resolution_table(result, s(i));
        end loop;
    end if;
    return result;
end function resolved;
```

The definition of the resolution function resolved.

Each of these is a closed subtype; that is, the result of resolving values in each case is a value within the range of the subtype. The subtype **X01Z** corresponds to the type **MVL4** we introduced in Figure 8-2 on pages 214–215.

The standard-logic package provides overloaded forms of the logical operators **and, nand, or, nor, xor, xnor** and **not** for standard-logic values and vectors, returning values in the range 'U', 'X', '0' or '1'. In addition, there are functions to convert between values of the full standard-logic type, the subtypes shown above and the predefined bit and bit-vector types. These are all listed in Appendix B.

VHDL-87

The VHDL-87 version of the standard-logic package does not provide the logical operator **xnor**, since **xnor** is not defined in VHDL-87.

8.3 Resolved Signals and Ports

In the previous discussion of resolved signals, we have limited ourselves to the simple case where a number of drivers or output ports of component instances drive a signal. Any input port connected to the resolved signal gets the final resolved value as the port value when a transaction is performed. We now look in more detail at the case of ports of mode **inout** being connected to a resolved signal. The question to answer here is, What value is seen by the input side of such a port? Is it the value driven by the component instance or the final value of the resolved signal connected to the port? In fact, it is the latter. An **inout** port models a connection in which the driver contributes to the associated signal's value, and the input side of the component senses the actual signal rather than using the driving value.

EXAMPLE

Some asynchronous bus protocols use a distributed synchronization mechanism based on a "wired-and" control signal. This is a single signal driven by each module using active-low open-collector or open-drain drivers and pulled up by the bus terminator. If a number of modules on the bus need to wait until all are ready to proceed with some operation, they use the control signal as follows. Initially, all modules drive the signal to the '0' state. When each is ready to proceed, it turns off its driver ('Z') and monitors the control signal. So long as any module is not yet ready, the signal remains at '0'. When all modules are ready, the bus terminator pulls the signal up to the '1' state. All modules sense this change and proceed with the operation.

Figure 8-9 shows an entity declaration for a bus module that has a port of the unresolved type **std_ulogic** for connection to such a synchronization control signal. The architecture body for a system comprising several such modules is also outlined. The control signal is pulled up by a concurrent signal assignment statement, which acts as a source with a constant driving value of 'H'. This is a value having a weak strength, which is overridden by any other source that drives '0'. It can pull the signal high only when all other sources drive 'Z'.

Figure 8-10 shows an outline of a behavioral architecture body for the bus module. Each instance initially drives its synchronization port with '0'. This value is passed up through the port and used as the contribution to the resolved signal from the entity instance. When an instance is ready to proceed with its operation, it changes its driving value to 'Z', modeling an open-collector or open-drain driver being turned off. The process then suspends until the value seen on the synchronization port changes to 'H'. If other instances are still driving '0', their contributions dominate, and the value of the signal stays '0'. When all other instances eventually change their contributions to 'Z', the value 'H' contributed by the pull-up statement dominates, and the value of the signal changes to 'H'. This value is passed back down through the ports of each instance, and the processes all resume.

FIGURE 8-9

```
library ieee;  use ieee.std_logic_1164.all;
entity bus_module is
    port ( synch : inout std_ulogic;  . . . );
end entity bus_module;
```
- -
```
architecture top_level of bus_based_system is
    signal synch_control : std_logic;
    . . .
begin
    synch_control_pull_up : synch_control <= 'H';
    bus_module_1 : entity work.bus_module(behavioral)
        port map ( synch => synch_control, . . . );
    bus_module_2 : entity work.bus_module(behavioral)
        port map ( synch => synch_control, . . . );

    . . .

end architecture top_level;
```

An entity declaration for a bus module that uses a "wired-and" synchronization signal, and an architecture body that instantiates the entity, connecting the synchronization port to a resolved signal.

FIGURE 8-10

```
architecture behavioral of bus_module is
begin
    behavior : process is
        . . .
    begin
        synch <= '0' after Tdelay_synch;
        . . .
        - - ready to start operation
        synch <= 'Z' after Tdelay_synch;
        wait until synch = 'H';
        - - proceed with operation
        . . .
    end process behavior;
end architecture behavioral;
```

An outline of a behavioral architecture body for a bus module, showing use of the synchronization control port.

Resolved Ports

Just as a signal declared with a signal declaration can be of a resolved subtype, so too can a port declared in an interface list of an entity. This is consistent with all that we have said about ports appearing just like signals to an architecture body. Thus if the architecture body contains a number of processes that must drive a port or a number of component instances that must connect outputs to a port, the port must be resolved. The final value driven by the resolved port is determined by resolving all of the sources within the architecture body. For example, we might declare an entity with a resolved port as follows:

```
library ieee;  use ieee.std_logic_1164.all;

entity IO_section is
    port ( data_ack : inout std_logic; . . . );
end entity IO_section;
```

The architecture body corresponding to this entity might instantiate a number of I/O controller components, each with their data acknowledge ports connected to the **data_ack** port of the entity. Each time any of the controllers updates its data acknowledge port, the standard-logic resolution function is invoked. It determines the driving value for the **data_ack** port by resolving the driving values from all controllers.

If it happens that the actual signal associated with a resolved port in an enclosing architecture body is itself a resolved signal, then the signal's resolution function will be called separately after the port's resolution function has determined the port's driving value. Note that the signal in the enclosing architecture body may use a different resolution function from the connected port, although in practice, most designs use the one function for resolution of all signals of a given subtype.

An extension of the above scenario is a design in which there are several levels of hierarchy, with a process nested at the deepest level generating a value to be passed out through resolved ports to a signal at the top level. At each level, a resolution function is called to determine the driving value of the port at that level. The value finally determined for the signal at the top level is called the *effective value* of the signal. It is passed back down the hierarchy of ports as the effective value of each **in** mode or **inout** mode port. This value is used on the input side of each port.

EXAMPLE

Figure 8-11 shows the hierarchical organization for a single-board computer system, consisting of a frame buffer for a video display, an input/output controller section, a CPU/memory section and a bus expansion block. These are all sources for the resolved data bus signal. The CPU/memory section in turn comprises a memory block and a CPU/cache block. Both of these act as sources for the data port, so it must be a resolved port. The cache has two sections, both of which act as sources for the data port of the CPU/cache block. Hence, this port must also be resolved.

Let us consider the case of one of the cache sections updating its data port. The new driving value is resolved with the current driving value from the other cache section to determine the driving value of the CPU/cache block data port. This result is then resolved with the current driving value of the memory block to determine the driving value of the CPU/memory section. Next, this driving value is resolved with the current driving values of the other top-level sections to determine the effective value of the data bus signal. The final step involves propagating this signal value back down the hierarchy for use as the effective value of each

FIGURE 8-11

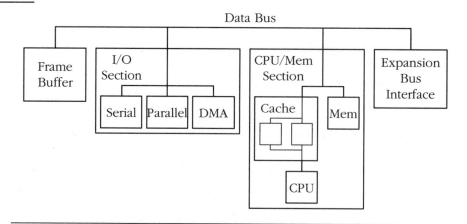

A hierarchical block diagram of a single-board computer system, showing the hierarchical connections of the resolved data bus ports to the data bus signal.

of the data ports. Thus, a module that reads the value of its data port will see the final resolved value of the data bus signal. This value is not necessarily the same as the driving value it contributes.

Driving Value Attribute

Since the value seen on a signal or on an **inout** mode port may be different from the value driven by a process, VHDL provides an attribute, 'driving_value, that allows the process to read the value it contributes to the prefix signal. For example, if a process has a driver for a resolved signal **s**, it may be driving **s** with the value 'Z' from a previously executed signal assignment statement, but the resolution function for **s** may have given it the value '0'. The process can refer to **s**'driving_value to retrieve the value 'Z'. Note that a process can only use this attribute to determine its own contribution to a signal; it cannot directly find out another process's contribution.

VHDL-87

The 'driving_value attribute is not provided in VHDL-87.

8.4 Resolved Signal Parameters

Let us now return to the topic of subprograms with signal parameters and see how they behave in the presence of resolved signals. Recall that when a procedure with an **out** mode signal parameter is called, the procedure is passed a reference to the caller's driver for the actual signal. Any signal assignment statements performed within the procedure body are actually performed on the caller's driver. If the actual signal parameter is a resolved signal, the values assigned by the procedure are used to resolve the signal value. No resolution takes place within the procedure. In fact, the procedure need not be aware that the actual signal is resolved.

In the case of an **in** mode signal parameter to a function or procedure, a reference to the actual signal parameter is passed when the subprogram is called, and the subprogram uses the actual value of the signal. If the signal is resolved, the subprogram sees the value determined after resolution. In the case of an **inout** signal parameter, a procedure is passed references to both the signal and its driver, and no resolution is performed internally to the procedure.

EXAMPLE

We can encapsulate the distributed synchronization protocol described in the example on page 223 in a set of procedures, each with a single signal parameter, as shown in Figure 8-12. Suppose a process uses a resolved signal **barrier** of subtype **std_logic** to synchronize with other processes. Figure 8-13 shows how the process might use the procedures to implement the protocol.

FIGURE 8-12

```
procedure init_synchronize ( signal synch : out std_logic ) is
begin
    synch <= '0';
end procedure init_synchronize;

procedure begin_synchronize ( signal synch : inout std_logic;
                              Tdelay : in delay_length := 0 fs ) is
begin
    synch <= 'Z' after Tdelay;
    wait until synch = 'H';
end procedure begin_synchronize;

procedure end_synchronize ( signal synch : inout std_logic;
                            Tdelay : in delay_length := 0 fs ) is
begin
    synch <= '0' after Tdelay;
    wait until synch = '0';
end procedure end_synchronize;
```

Three procedures that encapsulate the distributed synchronization operation.

The process has a driver for **barrier**, since the procedure calls associate the signal as an actual parameter with formal parameters of mode **out** and **inout**. A reference to this driver is passed to init_synchronize, which assigns the value '0' on behalf of the process. This value is used in the resolution of **barrier**. When the process is ready to start its synchronized operation, it calls begin_synchronize, passing references to its driver for **barrier** and to the actual signal itself. The procedure uses the driver to assign the value 'Z' on behalf of the process and then waits until the actual signal changes to 'H'. When the transaction on the driver matures, its value is resolved with other contributions from other processes and the result applied to the signal. This final value is used by the wait statement in the procedure to determine whether to resume the calling process. If the value is 'H', the process resumes,

FIGURE 8-13

```
synchronized_module : process is
    . . .
begin
    init_synchronize(barrier);
    . . .
    loop
        . . .
        begin_synchronize(barrier);
        . . .         -- perform operation, synchronized with other processes
        end_synchronize(barrier);
        . . .
    end loop;
end process synchronized_module;
```

An outline of a process that uses the distributed synchronization protocol procedures, with a resolved control signal barrier.

the procedure returns to the caller and the operation goes ahead. When the process completes the operation, it calls **end_synchronize** to reset **barrier** back to '0'.

Exercises

1. [❶ 8.1] Suppose there are four drivers connected to a resolved signal that uses the resolution function shown in Figure 8-1 on page 213. What is the resolved value of the signal if the four drivers contribute these values:

   (a) 'Z', '1', 'Z', 'Z'?

   (b) '0', 'Z', 'Z, '0'?

   (c) 'Z', '1', 'Z', '0'?

2. [❶ 8.1] Rewrite the following resolved signal declaration as a subtype declaration followed by a signal declaration using the subtype.

   signal synch_control : wired_and tri_state_logic := '0';

3. [❶ 8.1] What is the initial value of the following signal of the type **MVL4_logic** defined in Figure 8-2 on page 214? How is that value derived?

   signal int_req : MVL4_logic;

4. [❶ 8.1] Does the result of the resolution function defined in Figure 8-2 on pages 214–215 depend on the order of contributions from drivers in the array passed to the function?

5. [❶ 8.1] Suppose we define a resolved array subtype, **byte**, in the same way that the type **word** is defined in Figure 8-5 on page 216, but with eight elements in the array type instead of 32. We then declare a signal of type **byte** with three drivers. What is the resolved value of the signal if the three drivers contribute these values:

   (a) "ZZZZZZZZ", "ZZZZ0011", "ZZZZZZZZ"?

   (b) "XXXXZZZZ", "ZZZZZZZZ", "00000011"?

    (c) "00110011", "ZZZZZZZZ", "ZZZZ1111"?

6. [❶ 8.1] Suppose a signal is declared as

    signal data_bus : MVL4_logic_vector(0 **to** 15);

where MVL4_logic_vector is as described on page 218, and the following signal assignments are each executed in different processes:

    data_bus <= "ZZZZZZZZZZZZZZZZ";

    data_bus(0 to 7) <= "XXXXZZZZ";

    data_bus(8 to 15) <= "00111100";

What is the resolved signal value after all of the transactions have been performed?

7. [❶ 8.2] Suppose there are four drivers connected to a signal of type **std_logic**. What is the resolved value of the signal if the four drivers contribute these values:

    (a) 'Z', '0', 'Z', 'H'?

    (b) 'H', 'Z', 'W', '0'?

    (c) 'Z', 'W', 'L', 'H'?

    (d) 'U', '0', 'Z', '1'?

    (e) 'Z', 'Z', 'Z', '–'?

8. [❶ 8.3] Below is a timing diagram for the system with two bus modules using the wired-and synchronization signal described in Figure 8-9 on page 224. The diagram shows the driving values contributed by each of the bus modules to the synch_control signal. Complete the diagram by drawing the resolved waveform for synch_control. Indicate the times at which each bus module proceeds with its internal operation, as described in Figure 8-10.

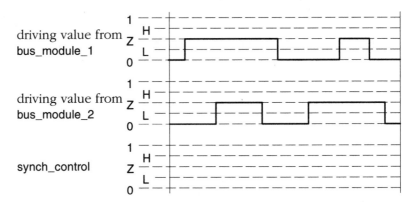

9. [❶ 8.3] Suppose all of the modules in the hierarchy of Figure 8-11 on page 226 use resolved ports for their data connections. If the Mem, Cache, Serial and DMA modules all update their data drivers in the same simulation cycle, how many times is the resolution function invoked to determine the final resolved values of the data signals?

10. [❶ 8.3] Suppose a process in a model drives a bidirectional port **synch_T** of type **std_logic**. Write a signal assignment statement that inverts the process's contribution to the port.

11. [❷ 8.1] Develop a model that includes two processes, each of which drives a signal of the type MVL4_logic described in Figure 8-2 on pages 214–215. Experiment with your simulator to see if it allows you to trace the invocation and execution of the resolution function.

12. [❷ 8.2] Develop a model of an inverter with an open-collector output of type std_logic, and a model of a pull-up resistor that drives its single std_logic port with the value 'H'. Test the models in a test bench that connects the outputs of a number of inverter instances to a signal of type std_logic, pulled up with a resistor instance. Verify that the circuit implements the active-low wired-or operation.

13. [❷ 8.2] Develop a behavioral model of an eight-bit-wide bidirectional transceiver, such as the 74245 family of components. The transceiver has two bidirectional data ports, a and b, an active-low output-enable port, oe_n, and a direction port, dir. When oe_n is low and dir is low, data is received from b to a. When oe_n is low and dir is high, data is transmitted from a to b. When oe_n is high, both a and b are high impedance. Assume a propagation delay of 5 ns for all output changes.

14. [❷ 8.2] Many combinatorial logic functions can be implemented in integrated circuits using pass transistors acting as switches. While a pass transistor is, in principle, a bidirectional device, for many circuits it is sufficient to model it as a unidirectional device. Develop a model of a unidirectional pass transistor switch, with an input port, an output port and an enable port, all of type std_logic. When the enable input is 'H' or '1', the input value is passed to the output, but with weak drive strength. When the enable input is 'L' or '0', the output is high impedance. If the enable input is at an unknown level, the output is unknown, except that its drive strength is weak.

15. [❸ 8.2] Develop a behavioral model of a tristate buffer with data input, data output and enable ports, all of type std_logic. The propagation time from data input to data output when the buffer is enabled is 4 ns. The turn-on delay from the enable port is 3 ns, and the turn-off delay is 3.5 ns. Use the buffer and any other necessary gate models in a structural model of the eight-bit transceiver described in Exercise 13.

16. [❸ 8.2] Use the unidirectional pass transistor model of Exercise 14 in a structural model of a four-input multiplexer. The multiplexer has select inputs s0 and s1. Pass transistors are used to construct the multiplexer as follows:

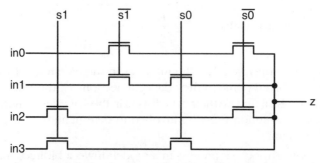

17. [❸ 8.2] Develop a model of a distributed priority arbiter for a shared bus in a multiprocessor computer system. Each bus requester has a request priority, *R*, between

0 and 31, with 0 indicating the most urgent request and 31 indicating no request. Priorities are binary-encoded using five-bit vectors, with bit 4 being the most-significant bit and bit 0 being the least-significant bit. The standard-logic values 'H' and '1' both represent the binary digit 1, and the standard-logic value '0' represents the binary digit 0. All requesters can drive and sense a five-bit arbitration bus, A, which is pulled up to 'H' by the bus terminator. The requesters each use A and their own priority to compute the minimum of all priorities by comparing the binary digits of priorities as follows. For each bit position i:

- if $(R_{4...i+1} = A_{4...i+1})$ and $(R_i = 0)$: drive A_i with '0' after T_{pd}
- if $(R_{4...i+1} \neq A_{4...i+1})$ or $(R_i = 1)$: drive A_i with 'Z' after T_{pd}

T_{pd} is the propagation delay between sensing a value on A and driving a resulting value on A. When the value on A has stabilized, it is the minimum of all request priorities. The requester with $R = A$ wins the arbitration. If you are not convinced that the distributed minimization scheme operates as required, trace its execution for various combinations of priority values.

Generic Constants

The models that we have used as examples in preceding chapters all have fixed behavior and structure. In many respects, this is a limitation, and we would like to be able to write more general, or *generic*, models. VHDL provides us with a mechanism, called *generics*, for writing parameterized models. We discuss generics in this chapter and show how they may be used to write families of models with varying behavior and structure.

9.1   Parameterizing Behavior

We can write a generic entity by including a *generic interface list* in its declaration that defines the *formal generic constants* that parameterize the entity. The extended syntax rule for entity declarations including generics is

> entity_declaration ⇐
>   **entity** identifier **is**
>   ⟦ **generic** (*generic*_interface_list) ; ⟧
>   ⟦ **port** (*port*_interface_list) ; ⟧
>   { entity_declarative_item }
>   ⟦ **begin**
>   { concurrent_assertion_statement
>   ∥ *passive*_concurrent_procedure_call_statement
>   ∥ *passive*_process_statement } ⟧
>   **end** ⟦ **entity** ⟧ ⟦ identifier ⟧ ;

The difference between this and the simpler rule we have seen before is the inclusion of the optional generic interface list before the port interface list. The generic interface list is like any other interface list, but with the restriction that we can only include constant-class objects, which must be of mode **in**. Since these are the defaults for a generic interface list, we can use a simplified syntax rule:

> *generic*_interface_list ⇐
>   (identifier { , ∘∘∘ } : subtype_indication ⟦ := expression ⟧)
>   { ; ∘∘∘ }

A simple example of an entity declaration including a generic interface list is

```
entity and2 is
    generic ( Tpd : time );
    port ( a, b : in bit;  y : out bit );
end entity and2;
```

This entity includes one generic constant, **Tpd**, of the predefined type **time**. The value of this generic constant may be used within the entity statements and any architecture body corresponding to the entity. In this example the intention is that the generic constant specify the propagation delay for the module, so the value should be used in a signal assignment statement as the delay. An architecture body that does this is

```
architecture simple of and2 is
begin

    and2_function :
        y <= a and b after Tpd;

end architecture simple;
```

The visibility of a generic constant extends from the end of the generic interface list to the end of the entity declaration and extends into any architecture body corresponding to the entity declaration.

A generic constant is given an actual value when the entity is used in a component instantiation statement. We do this by including a *generic map*, as shown by the extended syntax rule for component instantiations:

component_instantiation_statement ⟸
    *instantiation*_label :
        entity *entity*_name ⟦ (*architecture*_identifier) ⟧
        ⟦ **generic map** (*generic*_association_list) ⟧
        ⟦ **port map** (*port*_association_list) ⟧ ;

The generic association list is like other forms of association lists, but since generic constants are always of class constant, the actual arguments we supply must be expressions. Thus the simplified syntax rule for a generic association list is

*generic*_association_list ⟸
    (⟦ *generic*_name => ⟧ (expression ∥ **open**)) { , ... }

To illustrate this, let us look at a component instantiation statement that uses the and2 entity shown above:

gate1 : **entity** work.and2(simple)
    generic map (Tpd => 2 ns)
    port map (a => sig1, b => sig2, y => sig_out);

The generic map specifies that this instance of the **and2** module uses the value 2 ns for the generic constant **Tpd**; that is, the instance has a propagation delay of 2 ns. We might include another component instantiation statement using **and2** in the same design but with a different actual value for **Tpd** in its generic map, for example:

gate2 : **entity** work.and2(simple)
    generic map (Tpd => 3 ns)
    port map (a => a1, b => b1, y => sig1);

When the design is elaborated we have two processes, one corresponding to the instance **gate1** of and2, which uses the value 2 ns for **Tpd**, and another corresponding to the instance **gate2** of and2, which uses the value 3 ns.

EXAMPLE

As the syntax rule for the generic interface list shows, we may define a number of generic constants of different types and include default values for them. A more involved example is shown in Figure 9-1. In this example, the generic interface list includes a list of two generic constants that parameterize the propagation delay of the module and a Boolean generic constant, **debug**, with a default value of false. The intention of this last generic constant is to allow a design that instantiates this entity to activate some debugging operation. This operation might take the form of report statements within if statements that test the value of **debug**.

FIGURE 9-1

```
entity control_unit is
    generic ( Tpd_clk_out, Tpw_clk : delay_length;
            debug : boolean := false );
    port ( clk : in bit;
            ready : in bit;
            control1, control2 : out bit );
end entity control_unit;
```

An entity declaration for a block of sequential control logic, including generic constants that parameterize its behavior.

We have the same flexibility in writing a generic map as we have in other association lists. We can use positional association, named association or a combination of both. We can omit actual values for generic constants that have default expressions, or we may explicitly use the default value by writing the keyword **open** in the generic map. To illustrate these possibilities, here are three different ways of writing a generic map for the control_unit entity:

generic map (200 ps, 1500 ps, false)

generic map (Tpd_clk_out => 200 ps, Tpw_clk => 1500 ps)

generic map (200 ps, 1500 ps, debug => **open**)

EXAMPLE

Figure 9-2 shows the entity declaration and a behavioral architecture body for a D-flipflop. The model includes generic constants: **Tpd_clk_q** to specify the propagation delay from clock rising edge to output, **Tsu_d_clk** to specify the setup time of data before a clock edge and **Th_d_clk** to specify the hold time of data after a clock edge. The values of these generic constants are used in the architecture body.

FIGURE 9-2

```
entity D_flipflop is
    generic ( Tpd_clk_q, Tsu_d_clk, Th_d_clk : delay_length );
    port ( clk, d : in bit;  q : out bit );
end entity D_flipflop;
- - - - - - - - - - - - - - - - - - - - - - - - - - - - - - - - - - - - - - -
architecture basic of D_flipflop is
begin
    behavior : q <= d after Tpd_clk_q when clk = '1' and clk'event;
    check_setup : process is
    begin
        wait until clk = '1';
```

```
            assert d'last_event >= Tsu_d_clk
                report "setup violation";
        end process check_setup;

        check_hold : process is
        begin
            wait until clk'delayed(Th_d_clk) = '1';
            assert d'delayed'last_event >= Th_d_clk
                report "hold violation";
        end process check_hold;
    end architecture basic;
```

An entity and architecture body for a D-flipflop. The entity declaration includes generic constants for specifying timing characteristics. These are used within the architecture body.

The entity might be instantiated as follows, with actual values specified in the generic map for the generic constants:

```
    request_flipflop : entity work.D_flipflop(basic)
        generic map ( Tpd_clk_q => 4 ns,
                        Tsu_d_clk => 3 ns, Th_d_clk => 1 ns )
        port map ( clk => system_clock,
                        d => request, q => request_pending );
```

9.2   Parameterizing Structure

The second main use of generic constants in entities is to parameterize their structure. We can use the value of a generic constant to specify the size of an array port. To see why this is useful, let us look at an entity declaration for a register. A register entity that uses an unconstrained array type for its input and output ports can be declared as

```
    entity reg is
        port ( d : in bit_vector; q : out bit_vector; ... );
    end entity reg;
```

While this is a perfectly legal entity declaration, it does not include the constraint that the input and output ports d and q should be of the same size. Thus we could write a component instantiation as follows:

```
    signal small_data : bit_vector(0 to 7);
    signal large_data : bit_vector(0 to 15);
    . . .

    problem_reg : entity work.reg
        port map ( d => small_data, q => large_data, ... );
```

The model is analyzed and elaborated without the error being detected. It is only when the register tries to assign a small bit vector to a target bit vector of a larger size that the error is detected. We can avoid this problem by including a generic constant in the entity declaration to parameterize the size of the ports. We use the generic constant in constraints in the port declarations. To illustrate, here is the register entity declaration rewritten:

```
entity reg is
    generic ( width : positive );
    port ( d : in bit_vector(0 to width − 1);
           q : out bit_vector(0 to width − 1);
           . . . );
end entity reg;
```

In this declaration we require that the user of the register specify the desired port width for each instance. The entity then uses the width value as a constraint on both the input and output ports, rather than allowing their size to be determined by the signals associated with the ports. A component instantiation using this entity might appear as follows:

```
signal in_data, out_data : bit_vector(0 to bus_size − 1);
. . .
ok_reg : entity work.reg
    generic map ( width => bus_size )
    port map ( d => in_data,  q => out_data, . . . );
```

If the signals used as actual ports in the instantiation were of different sizes, the analyzer would signal the error early in the design process, making it easier to correct. As a matter of style, whenever the sizes of different array ports of an entity are related, generic constants should be considered to enforce the constraint.

EXAMPLE

A complete model for the register, including the entity declaration and an architecture body, is shown in Figure 9-3. The generic constant is used to constrain the widths of the data input and output ports in the entity declaration. It is also used in the architecture body to determine the size of the constant bit vector **zero**. This bit vector is the value assigned to the register output when it is reset, so it must be of the same size as the register port.

We can create instances of the register entity in a design, each possibly having different-sized ports. For example:

```
word_reg : entity work.reg(behavioral)
    generic map ( width => 32 )
    port map ( . . . );
```

creates an instance with 32-bit-wide ports. In the same design, we might include another instance, as follows:

```
subtype state_vector is bit_vector(1 to 5);

state_reg : entity work.reg(behavioral)
    generic map ( width => state_vector'length )
    port map ( . . . );
```

This register instance has five-bit-wide ports, wide enough to store values of the subtype state_vector.

FIGURE 9-3

```
entity reg is
    generic ( width : positive );
    port ( d : in  bit_vector(0 to width − 1);
           q : out  bit_vector(0 to width − 1);
           clk, reset : in bit );
end entity reg;

– – – – – – – – – – – – – – – – – – – – – – – – – – – – – – – – – – – – – –

architecture behavioral of reg is
begin

    behavior : process (clk, reset) is
        constant zero : bit_vector(0 to width − 1) := (others => '0');
    begin
        if reset = '1' then
            q <= zero;
        elsif clk'event and clk = '1' then
            q <= d;
        end if;
    end process behavior;

end architecture behavioral;
```

An entity and architecture body for a register with parameterized port size.

Exercises

1. [● 9.1]   Add to the following entity interface a generic clause defining generic constants Tpw_clk_h and Tpw_clk_l that specify the minimum clock pulse width timing. Both generic constants have a default value of 3 ns.

```
entity flipflop is
    port ( clk, d : in bit;  q, q_n : out bit );
end entity flipflop;
```

2. [● 9.1]   Write a component instantiation statement that instantiates the following entity from the current working library. The actual value for the generic constant should be 10 ns, and the clk signal should be associated with a signal called master_clk.

```
entity clock_generator is
    generic ( period : delay_length );
    port ( clk : out std_logic );
end entity clock_generator;
```

3. [❶ 9.2] Following is an incomplete entity interface that uses a generic constant to specify the sizes of the standard-logic vector input and output ports. Complete the interface by filling in the types of the ports.

```
entity adder is
    generic ( data_length : positive );
    port ( a, b : in . . .; sum : out . . . );
end entity adder;
```

4. [❶ 9.2] A system has an eight-bit data bus declared as

```
signal data_out : bit_vector(7 downto 0);
```

Write a component instantiation statement that instantiates the **reg** entity defined in Figure 9-3 to implement a four-bit control register. The register data input connects to the rightmost four bits of **data_out**, the **clk** input to io_write, the **reset** input to io_reset and the data output bits to control signals io_en, io_int_en, io_dir and io_mode.

5. [❷ 9.1] Develop a behavioral model of a D-latch with separate generic constants for specifying the following propagation delays:

- rising data input to rising data output,
- falling data input to falling data output,
- rising enable input to rising data output, and
- rising enable input to falling data output.

6. [❷ 9.1] Develop a behavioral model of a counter with output of type **natural** and clock and reset inputs of type **bit**. The counter has a Boolean generic constant, **trace_reset**. When this is true, the counter reports a trace message each time the reset input is activated.

7. [❷ 9.2] Develop a behavioral model of the adder described in Exercise 3.

8. [❷ 9.2] Develop a behavioral model of a multiplexer with n select inputs, 2^n data inputs and one data output.

9. [❸] Develop a behavioral model of a RAM with generic constants governing the read access time, minimum write time, the address port width and the data port width.

Components
and Configurations

In Chapter 5 we saw how to write entity declarations and architecture bodies that describe the structure of a system. Within an architecture body, we can write component instantiation statements that describe instances of an entity and connect signals to the ports of the instances. This simple approach to building a hierarchical design works well if we know in advance all the details of the entities we want to use. However, that is not always the case, especially in a large design project. In this chapter we introduce an alternative way of describing the hierarchical structure of a design that affords significantly more flexibility at the cost of a little more effort in managing the design.

10.1 Components

The first thing we need to do to describe an interconnection of subsystems in a design is to describe the different kinds of components used. We have seen how to do this by writing entity declarations for each of the subsystems. Each entity declaration is a separate design unit and has corresponding architecture bodies that describe implementations. An alternative approach is to write *component declarations* in the declarative part of an architecture body or package interface. We can then create *instances* of the components within the statement part of the architecture body.

Component Declarations

A component declaration simply specifies the external interface to the component in terms of generic constants and ports. We do not need to describe any corresponding implementation, since all we are interested in is how the component is connected in the current level of the design hierarchy. This makes the architecture completely self-contained, since it does not depend on any other library units except its corresponding entity interface. Let us look at the syntax rule that governs how we write a component declaration.

```
component_declaration ⇐
    component identifier [ is ]
        [ generic ( generic_interface_list ) ; ]
        [ port ( port_interface_list ) ; ]
    end component [ identifier ] ;
```

A simple example of a component declaration that follows this syntax rule is

```
component flipflop is
    generic ( Tprop, Tsetup, Thold : delay_length );
    port ( clk : in bit;  clr : in bit;  d : in bit;
           q : out bit );
end component flipflop;
```

This declaration defines a component type that represents a flipflop with clock, clear and data inputs, clk, clr and d, and a data output q. It also has generic constants for parameterizing the propagation delay, the data setup time and the data hold time.

Note the similarity between a component declaration and an entity declaration. This similarity is not accidental, since they both serve to define the external interface to a module. Although there is a very close relationship between components and entities, in fact they embody two different concepts. This may be a source of confusion to newcomers to VHDL. Nevertheless, the flexibility afforded by having the two different constructs is a powerful feature of VHDL, so we will work through it carefully in this section and try to make the distinction clear.

One way of thinking about the difference between an entity declaration and a component declaration is to think of the modules being defined as having different levels of "reality." An entity declaration defines a "real" module: something that ultimately will have a physical manifestation. For example, it may represent a circuit board in a rack, a packaged integrated circuit or a standard cell included in a piece of silicon. An entity declaration is a separate design unit that may be separately analyzed and placed into a design library. A component declaration, on the other hand, defines a "virtual," or "idealized," module that is included within an architecture body. It is as though we are saying, "For this architecture body, we assume there is a module as defined by this component declaration, since such a module meets our needs exactly." We specify the names, types and modes of the ports on the virtual module (the component) and proceed to lay out the structure of the design using this idealized view.

Of course, we do not make these assumptions about modules arbitrarily. One possibility is that we know what real modules are available and customize the virtual reality based on that knowledge. The advantage here is that the idealization cushions us from the irrelevant details of the real module, making the design easier to manage. Another possibility is that we are working "top down" and will later use the idealized module as the specification for a real module. Either way, eventually a link has to be made between an instance of a virtual component and a real entity so that the design can be constructed. In the rest of this section, we look at how to use components in an architecture body, then come back to the question of the binding between component instances and entities.

VHDL-87

The keyword **is** may not be included in the header of a component declaration, and the component name may not be repeated at the end of the declaration.

Component Instantiation

If a component declaration defines a kind of module, then a component instantiation specifies a usage of the module in a design. We have seen how we can instantiate an entity directly using a component instantiation statement within an architecture body. Let us now look at an alternative syntax rule that shows how we can instantiate a declared component:

component_instantiation_statement ⇐
    *instantiation*_label :
        ⟦ **component** ⟧ *component*_name
        ⟦ **generic map** (*generic*_association_list) ⟧
        ⟦ **port map** (*port*_association_list) ⟧ ;

This syntax rule shows us that we may simply name a component declared in the architecture body and, if required, provide actual values for the generic constants and actual signals to connect to the ports. The label is required to identify the component instance.

EXAMPLE

We can construct a four-bit register using flipflops and an and gate, similar to the example in Chapter 5. The entity declaration is shown at the top of Figure 10-1. The architecture body describing the structure of this register uses the flipflop component shown on page 242. Note that all we have done here is

FIGURE 10-1

```
entity reg4 is
    port ( clk, clr : in bit;  d : in bit_vector(0 to 3);
           q : out bit_vector(0 to 3) );
end entity reg4;

-------------------------------------------------

architecture struct of reg4 is

    component flipflop is
        generic ( Tprop, Tsetup, Thold : delay_length );
        port ( clk : in bit;  clr : in bit;  d : in bit;
               q : out bit );
    end component flipflop;

begin

    bit0 : component flipflop
        generic map ( Tprop => 2 ns, Tsetup => 2 ns, Thold => 1 ns )
        port map ( clk => clk, clr => clr, d => d(0), q => q(0) );

    bit1 : component flipflop
        generic map ( Tprop => 2 ns, Tsetup => 2 ns, Thold => 1 ns )
        port map ( clk => clk, clr => clr, d => d(1), q => q(1) );

    bit2 : component flipflop
        generic map ( Tprop => 2 ns, Tsetup => 2 ns, Thold => 1 ns )
        port map ( clk => clk, clr => clr, d => d(2), q => q(2) );

    bit3 : component flipflop
        generic map ( Tprop => 2 ns, Tsetup => 2 ns, Thold => 1 ns )
        port map ( clk => clk, clr => clr, d => d(3), q => q(3) );

end architecture struct;
```

An entity declaration and architecture body for a register using a component declaration for a flipflop.

specify the structure of this level of the design hierarchy, without having indicated how the flipflop is implemented. We will see how that may be done in the remainder of this chapter.

VHDL-87

The keyword **component** may not be included in a component instantiation statement in VHDL-87. The keyword is allowed in VHDL-93 to distinguish between instantiation of a component and direct instantiation of an entity. In VHDL-87, the only form of component instantiaton statement provided is instantiation of a declared component.

Packaging Components

Let us now turn to the issue of design management for large projects and see how we can make management of large libraries of entities easier using packages and components. Usually, work on a large design is partitioned among several designers, each responsible for implementing one or more entities that are used in the complete system. Each entity may need to have some associated types defined in a utility package, so that entity ports can be declared using those types. When the entity is used, other designers will need component declarations to instantiate components that will eventually be bound to the entity. It makes good sense to include a component declaration in the utility package, along with the types and other related items. This means that users of the entity do not need to rewrite the declarations, thus avoiding a potential source of errors and misunderstanding.

EXAMPLE

Suppose we are responsible for designing a serial interface cell for a microcontroller circuit. We can write a package specification that defines the interface to be used in the rest of the design, as outlined in Figure 10-2. The component declaration in this package corresponds to our entity declaration for the serial interface, shown in Figure 10-3. When other designers working on integrating the entire circuit need to instantiate the serial interface, they only need to import the items in the package, rather than rewriting all of the declarations. Figure 10-4 shows an outline of a design that does this.

FIGURE 10-2

```
library ieee;  use ieee.std_logic_1164.all;

package serial_interface_defs is

    subtype reg_address_vector is std_logic_vector(1 downto 0);

    constant status_reg_address : reg_address_vector := B"00";
    constant control_reg_address : reg_address_vector := B"01";
    constant rx_data_register : reg_address_vector := B"10";
    constant tx_data_register : reg_address_vector := B"11";

    subtype data_vector is std_logic_vector(7 downto 0);

    . . .        – – other useful declarations

    component serial_interface is
        port ( clock_phi1, clock_phi2 : in std_logic;
                serial_select : in std_logic;
                reg_address : in reg_address_vector;
                data : inout data_vector;
                interrupt_request : out std_logic;
                rx_serial_data : in std_logic;
                tx_serial_data : out std_logic );
    end component serial_interface;

end package serial_interface_defs;
```

An outline of a package declaration containing useful definitions for a serial interface.

FIGURE 10-3

```
library ieee;  use ieee.std_logic_1164.all;
use work.serial_interface_defs.all;
entity serial_interface is
     port ( clock_phi1, clock_phi2 : in std_logic;
            serial_select : in std_logic;
            reg_address : in reg_address_vector;
            data : inout data_vector;
            interrupt_request : out std_logic;
            rx_serial_data : in std_logic;
            tx_serial_data : out std_logic );
end entity serial_interface;
```

An entity declaration for the serial interface.

FIGURE 10-4

```
library ieee;  use ieee.std_logic_1164.all;
architecture structure of microcontroller is
     use work.serial_interface_defs.serial_interface;
     . . .         – – declarations of other components, signals, etc
begin
     serial_a : component serial_interface
          port map ( clock_phi1 => buffered_phi1,
                     clock_phi2 => buffered_phi2,
                     serial_select => serial_a_select,
                     reg_address => internal_addr(1 downto 0),
                     data => internal_data_bus,
                     interrupt_request => serial_a_int_req,
                     rx_serial_data => rx_data_a,
                     tx_serial_data => tx_data_a );
     . . .         – – other component instances
end architecture structure;
```

An outline of an architecture body that uses the serial interface definitions package and instantiates the component defined in it.

10.2 Configuring Component Instances

Once we have described the structure of one level of a design using components and component instantiations, we still need to flesh out the hierarchical implementation for each component instance. We can do this by writing a *configuration declaration* for the design. In it, we specify which real entity interface and corresponding architecture body should be used for each of the component instances. This is called *binding* the component instances to design entities. Note that we do not specify any binding information for a component instantiation statement that directly instantiates an entity, since the entity and architecture body are specified explicitly in the component instantiation statement. Thus our discussion in this section only applies to instantiations of declared components.

Basic Configuration Declarations

We start by looking at a simplified set of syntax rules for configuration declarations, as the full set of rules is rather complicated. The simplest case arises when the entities to which component instances are bound are implemented with behavioral architectures. In this case, there is only one level of the hierarchy to flesh out. The simplified syntax rules are

configuration_declaration ⇐
    configuration identifier **of** *entity*_name **is**
        for *architecture*_name
            { **for** component_specification
                binding_indication ;
                end for ; }
        end for ;
        end ⟦ **configuration** ⟧ ⟦ identifier ⟧ ;
component_specification ⇐
    (*instantiation*_label { , ... } ‖ **others** ‖ **all**) : *component*_name
binding_indication ⇐ **use entity** *entity*_name ⟦ (*architecture*_identifier) ⟧

The identifier given in the configuration declaration identifies this particular specification for fleshing out the hierarchy of the named entity. There may be other configuration declarations, with different names, for the same entity. Within the configuration declaration we write the name of the particular architecture body to work with (included after the first **for** keyword), since there may be several corresponding to the entity. We then include the binding information for each component instance within the architecture body. The syntax rule shows that we can identify a component instance by its label and its component name, as used in the component instantiation in the architecture body. We bind it by specifying an entity name and a corresponding architecture body name. For example, we might bind instances bit0 and bit1 of the component flipflop as follows:

for bit0, bit1 : flipflop
    use entity work.edge_triggered_Dff(basic);
end for;

This indicates that the instances are each to be bound to the design entity edge_triggered_Dff, found in the current working library, and that the architecture body basic corresponding to that entity should be used as the implementation of the instances.

Note that since we can identify each component instance individually, we have the opportunity to bind different instances of a given component to different entity/architecture pairs. After we have specified bindings for some of the instances in a design, we can use the keyword **others** to bind any remaining instances of a given component type to a given entity/architecture pair. Alternatively, if all instances of a particular component type are to have the same binding, we can use the keyword **all** instead of naming individual instances. The syntax rules also show that the architecture name corresponding to the entity is optional. If it is omitted, a default binding takes place when the design is elaborated for simulation or synthesis. The component instance is bound to whichever architecture body for the named entity has been most recently analyzed at the time of elaboration.

A configuration declaration is a primary design unit, and as such, may be separately analyzed and placed into the working design library as a library unit. If it contains sufficient binding information so that the full design hierarchy is fleshed out down to behavioral architectures, the configuration may be used as the target unit of a simulation. The design is elaborated by substituting instances of the specified architecture bodies for bound component instances in the way described in Section 5.5. The only difference is that when component declarations are instantiated, the configuration must be consulted to find the appropriate architecture body to substitute.

EXAMPLE

Let us look at a sample configuration declaration that binds the component instances in the four-bit register of Figure 10-1. Suppose we have a resource library for a project, **star_lib**, that contains the basic design entities that we need to use. Our configuration declaration might be written as shown in Figure 10-5.

FIGURE 10-5

```
library star_lib;
use star_lib.edge_triggered_Dff;
configuration reg4_gate_level of reg4 is
    for struct  -- architecture of reg4
        for bit0 : flipflop
            use entity edge_triggered_Dff(hi_fanout);
        end for;
        for others : flipflop
            use entity edge_triggered_Dff(basic);
        end for;
    end for;  -- end of architecture struct
end configuration reg4_gate_level;
```

A configuration declaration for a four-bit register model.

The library clause preceding the design unit is required to locate the resource library containing the entities we need. The use clause following it makes the entity names we require directly visible in the configuration declaration. The configuration is called **reg4_gate_level** and selects the architecture **struct** of the **reg4** entity. Within this architecture, we single out the instance **bit0** of the **flipflop** component and bind it to the entity **edge_triggered_Dff** with architecture **hi_fanout**. This shows how we can give special treatment to particular component instances when configuring bindings. We bind all remaining instances of the **flipflop** component to the **edge_triggered_Dff** entity using the **basic** architecture.

VHDL-87

The keyword **configuration** may not be included at the end of a configuration declaration in VHDL-87.

Configuring Multiple Levels of Hierarchy

In the previous section, we saw how to write a configuration declaration for a design in which the instantiated components are bound to behavioral architecture bodies. Most realistic designs, however, have deeper hierarchical structure. The components at the top level have architecture bodies that, in turn, contain component instances that must be configured. The architecture bodies bound to these second-level components may also contain component instances, and so on. In order to deal with configuring these more complex hierarchies, we need to use an alternative form of binding indication in the configuration declaration. The alternative syntax rule is

binding_indication $\Leftarrow$ **use configuration** *configuration*_name

This form of binding indication for a component instance allows us to bind to a preconfigured entity/architecture pair simply by naming the configuration declaration for the entity. For example, a component instance of **reg4** with the label **flag_reg** might be bound in a configuration declaration as follows:

```
for flag_reg : reg4
    use configuration work.reg4_gate_level;
end for;
```

EXAMPLE

In Chapter 5 we looked at a two-digit decimal counter, implemented using four-bit registers. We assume that the type **digit** is defined as follows in a package named **counter_types**:

```
subtype digit is bit_vector(3 downto 0);
```

The entity declaration for **counter** is shown at the top of Figure 10-6. Now that we have seen how to use component declarations, we can rewrite the architecture body using component declarations for the registers, as shown at the bottom of Figure 10-6. We can configure this implementation of the counter with the configuration declaration shown in Figure 10-7. This configuration specifies that each instance of the **digit_register** component is bound using the information in the configuration declaration named **reg4_gate_level** in the current design library, shown in Figure 10-5. That configuration in turn specifies the entity to use (**reg4**), a corresponding architecture body (**struct**) and the bindings for each component instance in that architecture body. Thus the two configuration declarations combine to fully configure the design hierarchy down to the process level.

FIGURE 10-6

```
use work.counter_types.digit;
entity counter is
    port ( clk, clr : in bit;
            q0, q1 : out digit );
end entity counter;
```

– –

```
architecture registered of counter is
    component digit_register is
        port ( clk, clr : in bit;
                d : in digit;
                q : out digit );
    end component digit_register;
    signal current_val0, current_val1, next_val0, next_val1 : digit;
begin
    val0_reg : component digit_register
        port map ( clk => clk, clr => clr, d => next_val0,
                    q => current_val0 );
    val1_reg : component digit_register
        port map ( clk => clk, clr => clr, d => next_val1,
                    q => current_val1 );
    – – other component instances
    . . .
end architecture registered;
```

An entity declaration and architecture body for a two-digit counter, using a component representing a four-bit register.

FIGURE 10-7

```
configuration counter_down_to_gate_level of counter is
    for registered
        for all : digit_register
            use configuration work.reg4_gate_level;
        end for;
        . . .        – – bindings for other component instances
    end for;  – – end of architecture registered
end configuration counter_down_to_gate_level;
```

A configuration declaration for the decimal counter.

The example above shows how we can use separate configuration declarations for each level of a design hierarchy. As a matter of style this is good practice, since it prevents the configuration declarations themselves from becoming too complex. The alternative approach is to configure an entity and its hierarchy fully within the one configuration declaration. We look at how this may be done, as some models from other designers may take this approach. While this approach is valid VHDL, we recommend the practice of splitting up the configuration information into separate configuration declarations corresponding to the entities used in the design hierarchy.

To see how to configure multiple levels within one declaration, we need to look at a more complex form of syntax rule for configuration declarations. In fact, we need to split the rule into two parts, so that we can write a recursive syntax rule.

configuration_declaration ⇐
    configuration identifier **of** *entity*_name **is**
        block_configuration
    end ⟦ **configuration** ⟧ ⟦ identifier ⟧ ;

block_configuration ⇐
    for *architecture*_name
        { **for** component_specification
            binding_indication ;
            ⟦ block_configuration ⟧
          end for ; }
    end for ;

The rule for a block configuration indicates how to write the configuration information for an architecture body and its inner component instances. (The reason for the name "block configuration" in the second rule is that it applies to block statements as well as architecture bodies. Block statements are considered an advanced topic and are not discussed in this book.) Note that we have included an extra part after the binding indication for a component instance. If the architecture that we bind to an instance also contains component instances, we can nest further configuration information for that architecture inside the enclosing block configuration.

EXAMPLE

We can write a configuration declaration equivalent to that in Figure 10-7 but containing all of the configuration information for the entire hierarchy, as shown in Figure 10-8. The difference between this configuration declaration and the one in Figure 10-7 is that the binding indication for instances of digit_register directly refers to the entity reg4 and the architecture body struct, rather than using a separate configuration for the entity. The configuration then includes all of the binding information for component instances within struct. This relatively simple example shows how difficult it can be to read nested configuration declarations. Separate configuration declarations are easier to understand and provide more flexibility for managing alternative compositions of a design hierarchy.

FIGURE 10-8

```
library star_lib;
use star_lib.edge_triggered_Dff;
configuration full of counter is
    for registered  – – architecture of counter
        for all : digit_register
            use entity work.reg4(struct);
            for struct  – – architecture of reg4
                for bit0 : flipflop
                    use entity edge_triggered_Dff(hi_fanout);
                end for;
                for others : flipflop
                    use entity edge_triggered_Dff(basic);
                end for;
            end for;  – – end of architecture struct
        end for;
        . . .        – – bindings for other component instances
    end for;  – – end of architecture registered
end configuration full;
```

An alternate configuration declaration for the decimal counter.

Direct Instantiation of Configured Entities

As we have seen, a configuration declaration specifies the design hierarchy for a design entity. We can make direct use of a fully configured design entity within an architecture body by writing a component instantiation statement that directly names the configuration. The alternative syntax rule for component instantiation statements that expresses this possibility is

component_instantiation_statement ⟸
    *instantiation*_label :
        configuration *configuration*_name
        ⟦ **generic map** (*generic*_association_list) ⟧
        ⟦ **port map** (*port*_association_list) ⟧ ;

The configuration named in the statement includes a specification of an entity and a corresponding architecture body to use. We can include generic and port maps in the component instantiation to provide actual values for any generic constants of the entity and actual signals to connect to the ports of the entity. This is much like instantiating the entity directly, but with all of the configuration information for its implementation included.

EXAMPLE

Figure 10-9 shows an outline of an architecture body that directly instantiates the two-digit decimal counter entity. The component instantiation statement labeled **seconds** refers to the configuration counter_down_to_gate_level, shown in Figure 10-7. That configuration, in turn, specifies the counter entity and architecture to use.

FIGURE 10-9

```
architecture top_level of alarm_clock is
    use work.counter_types.digit;
    signal reset_to_midnight, seconds_clk : bit;
    signal seconds_units, seconds_tens : digit;
    . . .
begin
    seconds : configuration work.counter_down_to_gate_level
        port map ( clk => seconds_clk, clr => reset_to_midnight,
                   q0 => seconds_units, q1 => seconds_tens );

    . . .
end architecture top_level;
```

An outline of an architecture body that directly instantiates the configured decimal counter entity.

VHDL-87

VHDL-87 does not allow direct instantiation of configured entities. Instead, we must declare a component, instantiate the component, and write a separate configuration declaration that binds the instance to the configured entity.

Generic and Port Maps in Configurations

We now turn to a very powerful and important aspect of component configurations: the inclusion of generic maps and port maps in the binding indications. This facility provides a great deal of flexibility when binding component instances to design entities. However, the ideas behind the facility are somewhat difficult to grasp on first encounter, so we will work through them carefully. First, let us look at an extended syntax rule for a binding indication that shows how generic and port maps can be included:

binding_indication ⇐
    use ⟦ **entity** *entity*_name ⟦ (*architecture*_identifier) ⟧
            ⟦ **configuration** *configuration*_name)
    ⟦ **generic map** (*generic*_association_list) ⟧
    ⟦ **port map** (*port*_association_list) ⟧

This rule indicates that after specifying the entity to which to bind (either directly or by naming a configuration), we may include a generic map or a port map or both. We show how this facility may be used by starting with some simple examples illustrating the more common uses. We then proceed to the general case.

One of the most important uses of this facility is to separate the specification of generic constants used for timing from the structure of a design. We can write component declarations in a structural description without including generic constants for timing. Later, when we bind each component instance to an entity in a configuration declaration, we can specify the timing values by supplying actual values for the generic constants of the bound entities.

EXAMPLE

Suppose we are designing an integrated circuit for a controller, and we wish to use the register whose entity declaration is shown in Figure 10-10. We can write a component declaration for the register without including the generic constants used for timing, as shown in the architecture body outlined in Figure 10-11. This component represents a virtual module that has all of the structural characteristics we need, but ignores timing. The component instantiation statement specifies a value for the port width generic constant, but does not specify any timing parameters.

FIGURE 10-10

```
library ieee;  use ieee.std_logic_1164.all;

entity reg is
    generic ( t_setup, t_hold, t_pd : delay_length;
              width : positive );
    port ( clock : in std_logic;
           data_in : in std_logic_vector(0 to width - 1);
           data_out : out std_logic_vector(0 to width - 1) );
end entity reg;
```

An entity declaration for a register, including generic constants for timing and port width.

FIGURE 10-11

```
architecture structural of controller is

    component reg is
        generic ( width : positive );
        port ( clock : in std_logic;
               data_in : in std_logic_vector(0 to width - 1);
               data_out : out std_logic_vector(0 to width - 1) );
    end component reg;

    . . .

begin
```

```
        state_reg : component reg
            generic map ( width => state_type'length )
            port map ( clock => clock_phase1,
                        data_in => next_state,
                        data_out => current_state );

    . . .

end architecture structural;
```

An outline of a structural architecture body of a controller design, using an idealized representation of the register module.

Since we are operating in the real world, we cannot ignore timing forever. Ultimately the values for the timing parameters will be determined from the physical layout of the integrated circuit. Meanwhile, during the design phase, we can use estimates for their values. When we write a configuration declaration for our design, we can configure the component instance as shown in Figure 10-12, supplying the estimates in a generic map. Note that we also need to specify a value for the **width** generic of the bound entity. In this example, we supply the value of the **width** generic of the component instance. We discuss this in more detail on page 257.

FIGURE 10-12

```
configuration controller_with_timing of controller is

    for structural

        for state_reg : reg
            use entity work.reg(gate_level)
            generic map ( t_setup => 200 ps, t_hold => 150 ps,
                            t_pd => 150 ps, width => width );
        end for;

        . . .

    end for;
end configuration controller_with_timing;
```

A configuration for the controller circuit, supplying values for the timing parameters of the register instance.

When we simulate the design, the estimated values for the generic constants are used by the real design entity to which the component instance is bound. Later, when the integrated circuit has been laid out, we can substitute, or *back annotate*, the actual timing values in the configuration declaration without having to modify the architecture body of the model. We can then resimulate to obtain test vectors for the circuit that take account of the real timing.

Another important use of generic and port maps in a configuration declaration arises when the entity to which we want to bind a component instance has different names for generic constants and ports. The maps in the binding indication can be used to make the link between component generics and ports on the one hand, and entity generics and ports on the other. Furthermore, the entity may have additional generics or ports beyond those of the component instance. In this case, the maps can be used to associate actual values or signals from the architecture body with the additional generics or ports.

EXAMPLE

Suppose we need to use a two-input-to-four-output decoder in a design, as shown in the outline of an architecture body in Figure 10-13. The component declaration for the decoder represents a virtual module that meets our needs exactly.

FIGURE 10-13

```
architecture structure of computer_system is
    component decoder_2_to_4 is
        generic ( prop_delay : delay_length );
        port ( in0, in1 : in bit;
                 out0, out1, out2, out3 : out bit );
    end component decoder_2_to_4;
    . . .
begin
    interface_decoder : component decoder_2_to_4
        generic map ( prop_delay => 4 ns )
        port map ( in0 => addr(4), in1 => addr(5),
                     out0 => interface_a_select, out1 => interface_b_select,
                     out2 => interface_c_select, out3 => interface_d_select );
    . . .
end architecture structure;
```

An outline of an architecture body for a computer system, using a four-output decoder component.

Now suppose we check in our library of entities for a real module to use for this instance and find a three-input-to-eight-output decoder. The entity declaration is shown in Figure 10-14. We could make use of this entity in our design if we could adapt to the different generic and port names and tie the unused ports to appropriate values. The configuration declaration in Figure 10-15 shows how this may be done. The generic map in the binding indication specifies the correspondence between entity generics and component generics. In this case, the component generic **prop_delay** is to be used for both entity generics. The port map in the binding indication similarly specifies which entity ports correspond to which component ports. Where the entity has extra ports, we can specify how those ports are to be connected. In this design, **s2** is tied to '0', **enable** is tied to '1' and the remaining ports are left unassociated (specified by the keyword **open**).

FIGURE 10-14

```
entity decoder_3_to_8 is
    generic ( Tpd_01, Tpd_10 : delay_length );
    port ( s0, s1, s2 : in bit;
            enable : in bit;
            y0, y1, y2, y3, y4, y5, y6, y7 : out bit );
end entity decoder_3_to_8;
```

An entity declaration for the real decoder module.

FIGURE 10-15

```
configuration computer_structure of computer_system is
    for structure
        for interface_decoder : decoder_2_to_4
            use entity work.decoder_3_to_8(basic)
            generic map ( Tpd_01 => prop_delay, Tpd_10 => prop_delay )
            port map ( s0 => in0, s1 => in1, s2 => '0',
                        enable => '1',
                        y0 => out0, y1 => out1, y2 => out2, y3 => out3,
                        y4 => open, y5 => open, y6 => open, y7 => open );
        end for;

        . . .

    end for;
end configuration computer_structure;
```

A configuration declaration for the computer system design, showing how the decoder component instance is bound to the real decoder entity.

The two preceding examples illustrate the most common uses of generic maps and port maps in configuration declarations. We now look at the general mechanism that underlies these examples, so that we can understand its use in more complex cases. We use the terms *local generics* and *local ports* to refer to the generics and ports of a component. Also, in keeping with previous discussions, we use the terms *formal generics* and *formal ports* to refer to the generics and ports of the entity to which the instance is bound.

When we write a component instantiation statement with a generic map and a port map, these maps associate actual values and signals with the *local* generics and ports of the component instance. Recall that the component is just a virtual module used as a template for a real module, so at this stage we have just made connections to the template. Next, we write a configuration declaration that binds the component instance to a real entity. The generic and port maps in the binding indication associate actual values and signals with the *formal* generics and ports of the entity. These actual values and signals may be the locals from the component instance, or they may be values and signals from the architecture body containing the component instance.

Figure 10-16 illustrates the mappings. It is this two-stage association mechanism that makes configurations so powerful in mapping a design to real modules.

Figure 10-16 shows that the actual values and signals supplied in the configuration declaration may be local generics or ports from the component instance. This is the case for the formal generics Tpd_01 and Tpd_10 and for the formal ports s0, s1, y0, y1, y2 and y3 in Figure 10-15. Every local generic and port of the component instance must be associated with a formal generic or port, respectively; otherwise the design is in error. The figure also shows that the configuration declaration may supply values or signals from the architecture body. Furthermore, they may be any other value or signal visible at the point of the component instantiation statement, such as the literals '0' and '1' shown in the example. Note that while it is legal to associate a signal in the architecture body with a formal port of the entity, it is not good practice to do so. This effectively modifies the structure of the circuit, making the overall design much more difficult to understand and manage. For example, in the configuration in Figure 10-15, had we associated the formal port s2 with the signal addr(6) instead of the literal value '0', the operation of the circuit would be substantially altered.

FIGURE 10-16

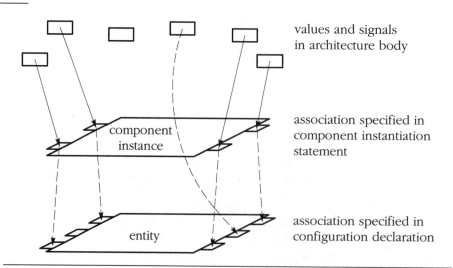

values and signals
in architecture body

association specified in
component instantiation
statement

association specified in
configuration declaration

The generic and port maps in the component instantiation and the configuration declaration define a two-stage association. Values and signals in the architecture body are associated, via the local generics and ports, with the formal generics and ports of the bound entity.

The preceding examples show how we can use generic and port maps in binding indications to deal with differences between the component and the entity in the number and names of generics and ports. However, if the component and entity have similar interfaces, we can rely on a *default binding* rule. This rule is used automatically if we omit the generic map or the port map in a binding indication, as we did in the earlier examples in this section. The default rule causes each local generic or port of the component to be associated with a formal generic or port of the same name in the entity interface. If the entity interface includes further formal generics or ports, they remain open. If the entity does not include a formal with the same name as one of the locals, the design is in error. So, for example, if we declare a component as

```
component nand3 is
    port ( a, b, c : in bit := '1';  y : out bit );
end component nand3;
```

and instantiate it as

```
gate1 : component nand3
    port map ( a => s1, b => s2, c => open, y => s3 );
```

then attempt to bind to an entity declared as

```
entity nand2 is
    port ( a, b : in bit := '1';  y : out bit );
end entity nand2;
```

with a component configuration

```
for gate1 : nand3
    use entity work.nand2(basic);
end for;
```

an error occurs. The reason for the error is that there is no formal port named **c** to associate with the local port of that name. The default rule requires that such a correspondence be found, even though the local port is unconnected in the architecture body.

Deferred Component Binding

We have seen that we can specify the binding for a component instance either by naming an entity and a corresponding architecture body, or by naming a configuration. A third option is to leave the component instance unbound and to defer binding it until later in the design cycle. The syntax rule for a binding indication that expresses this option is

binding_indication $\Leftarrow$ **use open**

If we use this form of binding indication to leave a component instance unbound, we cannot include a generic map or port map. This makes sense: since there is no entity, there are no formal generics or ports with which to associate actual values or signals.

A scenario in which we may wish to defer binding arises in complex designs that can be partially simulated before all subsystems are complete. We can write an architecture body for the system, including component declarations and instances as placeholders for the subsystems. Initially, we write a configuration declaration that defers bindings of the subsystems. Then, as the design of each subsystem is completed, the corresponding component configuration is updated to bind to the new entity. At intermediate stages it may be possible to simulate the system with some of the components unbound. The effect of the deferred bindings is simply to leave the corresponding ports unassociated when the design is elaborated. Thus the inputs to the unbound modules are not used, and the outputs remain undriven.

EXAMPLE

Figure 10-17 shows an outline of a structural architecture for a single-board computer system. The design includes all of the components needed to construct the system, including a CPU, main memory and a serial interface. However, if we have not yet designed an entity and architecture body for the serial interface, we cannot bind the component instance for the interface. Instead, we must leave it unbound, as shown in the configuration declaration in Figure 10-18. We can proceed to simulate the design, using the implementations of the CPU and main memory, provided we do not try to exercise the serial interface. If the processor were to try to access registers in the serial interface, it would get no response. Since there is no entity bound to the component instance representing the interface, there is nothing to drive the data or other signals connected to the instance.

FIGURE 10-17

```
architecture structural of single_board_computer is
    . . .        - - type and signal declarations
    component processor is
        port ( clk : in bit;  a_d : inout word; . . . );
    end component processor;
    component memory is
        port ( addr : in bit_vector(25 downto 0); . . . );
    end component memory;
    component serial_interface is
        port ( clk : in bit;  address : in bit_vector(3 downto 0); . . . );
    end component serial_interface;
begin
    cpu : component processor
        port map ( clk => sys_clk, a_d => cpu_a_d, . . . );
    main_memory : component memory
        port map ( addr => latched_addr(25 downto 0), . . . );
    serial_interface_a : component serial_interface
        port map ( clk => sys_clk, address => latched_addr(3 downto 0), . . . );
    . . .
end architecture structural;
```

An outline of an architecture body for a single-board computer, including component declarations and instances for the CPU, main memory and a serial interface controller.

FIGURE 10-18

```
library chips;
configuration intermediate of single_board_computer is
    for structural
```

```
            for cpu : processor
                use entity chips.XYZ3000_cpu(full_function)
                    port map ( clock => clk, addr_data => a_d, . . . );
            end for;
            for main_memory : memory
                use entity work.memory_array(behavioral);
            end for;
            for all : serial_interface
                use open;
            end for;

            . . .

        end for;
end configuration intermediate;
```

A configuration declaration for the single-board computer, in which the serial interface is left unbound.

Exercises

1. [❶ 10.1] List some of the differences between an entity declaration and a component declaration.

2. [❶ 10.1] Write a component declaration for a binary magnitude comparitor, with two standard-logic vector data inputs, **a** and **b**, whose length is specified by a generic constant, and two standard-logic outputs indicating whether **a** = **b** and **a** < **b**. The component also includes a generic constant for the propagation delay.

3. [❶ 10.1] Write a component instantiation statement that instantiates the magnitude comparitor described in Exercise 2. The data inputs are connected to signals current_position and upper_limit, the output indicating whether **a** < **b** is connected to position_ok and the remaining output is open. The propagation delay of the instance is 12 ns.

4. [❶ 10.1] Write a package declaration that defines a subtype of natural numbers representable in eight bits and a component declaration for an adder that adds values of the subtype.

5. [❶ 10.2] Suppose we have an architecture body for a digital filter, outlined as follows:

```
architecture register_transfer of digital_filter is
    . . .
    component multiplier is
        port ( . . . );
    end component multiplier;
begin
    coeff_1_multiplier : component multiplier
        port map ( . . . );
    . . .
end architecture register_transfer;
```

Write a configuration declaration that binds the multiplier component instance to a multiplier entity called fixed_point_mult from the library dsp_lib, using the architecture algorithmic.

6. [❶ 10.2] Suppose the library dsp_lib referred to in Exercise 5 includes a configuration of the fixed_point_mult entity called fixed_point_mult_std_cell. Write an alternative configuration declaration for the filter described in Exercise 5, binding the multiplier instance using the fixed_point_mult_std_cell configuration.

7. [❶ 10.2] Modify the outline of the filter architecture body described in Exercise 5 to directly instantiate the fixed_point_mult_std_cell configuration described in Exercise 6, rather than using the multiplier component.

8. [❶ 10.2] Suppose we declare and instantiate a multiplexer component in an architecture body as follows:

```
component multiplexer is
    port ( s, d0, d1 : in bit; z : out bit );
end component multiplexer;

serial_data_mux : component multiplexer
    port map ( s => serial_source_select,
               d0 => rx_data_0, d1 => rx_data_1,
               z => internal_rx_data );
```

Write a binding indication that binds the component instance to the following entity in the current working library, using the most recently analyzed architecture and specifying a value of 3.5 ns for the propagation delay.

```
entity multiplexer is
    generic ( Tpd : delay_length := 3 ns );
    port ( s, d0, d1 : in bit; z : out bit );
end entity multiplexer;
```

9. [❶ 10.2] Draw a diagram, based on Figure 10-16 on page 258, that shows the mapping between entity ports and generics, component ports and generics and other values in the configured computer system model of Figure 10-15 on page 257.

10. [❶ 10.2] Suppose we have an entity nand4 with the following interface in a library gate_lib:

```
entity nand4 is
    generic ( Tpd_01, Tpd_10 : delay_length := 2 ns );
    port ( a, b, c, d : in bit := '1';  y : out bit );
end entity nand4;
```

We bind the entity to the component instance gate1 described on page 259 using the following component configuration:

```
for gate1 : nand3
    use entity get_lib.nand4(basic);
end for;
```

Write the generic and port maps that comprise the default binding indication used in this configuration.

11. [❷ 10.1] Develop a structural model of a 32-bit bidirectional transceiver, implemented using a component based on the eight-bit transceiver described in Exercise 13 in Chapter 8.

12. [❷ 10.1] Develop a structural model for an eight-bit serial-in/parallel-out shift register, assuming you have available a four-bit serial-in/parallel-out shift register. Include a component declaration for the four-bit register, and instantiate it as required for the eight-bit register. The four-bit register has a positive-edge-triggered clock input, an active-low asynchronous reset input, a serial data input and four parallel data outputs.

13. [❷ 10.1] Develop a package of component declarations for two-input gates and an inverter, corresponding to the logical operators in VHDL. Each component has ports of type **bit** and generic constants for rising output and falling output propagation delays.

14. [❷ 10.2] Develop a configuration declaration for the 32-bit transceiver described in Exercise 11 that binds each instance of the eight-bit transceiver component to the eight-bit transceiver entity.

15. [❷ 10.2] Develop a behavioral model of a four-bit shift register that implements the component interface described in Exercise 12. Write a configuration declaration for the eight-bit shift register, binding the component instances to the four-bit shift register entity.

16. [❷ 10.2] Suppose we wish to use an XYZ1234A serial interface controller in the microcontroller described in Figure 10-4. The entity interface for the XYZ1234A is

```
entity XYZ1234A is
    generic ( T_phi_out, T_d_z : delay_length;
              debug_trace : boolean := false );
    port ( phi1, phi2 : in std_logic;              -- 2 phase clock
           cs : in std_logic;                      -- chip select
           a : in std_logic_vector(1 downto 0);    -- address
           d : inout std_logic_vector(1 downto 0); -- data
           int_req : out std_logic;                --  interrupt
           rx_d : in std_logic;                    -- rx serial data
           tx_d : out std_logic );                 -- tx serial data
end entity XYZ1234A;
```

Write a configuration declaration that binds the serial_interface component instance to the XYZ1234A entity, using the most recently compiled architecture, setting both timing generics to 6 ns and using the default value for the debug_trace generic.

17. [❷ 10.1/10.2] Use the package described in Exercise 13 to develop a structural model of a full adder, described by the Boolean equations

$$S = (A \oplus B) \oplus C_{in}$$
$$C_{out} = A \cdot B + (A \oplus B) \cdot C_{in}$$

Write behavioral models of entities corresponding to each of the gate components, and a configuration declaration that binds each component instance in the full adder to the appropriate gate entity.

18. [❸ 10.2] Develop a structural model of a four-bit adder using instances of a full-adder component. Write a configuration declaration that binds each instance of the full-adder component, using the configuration declaration described in Exercise 17. For comparison, write an alternative configuration declaration that fully configures the four-bit adder hierarchy without using the configuration declaration described in Exercise 17.

19. [❸ 10.2] Develop a behavioral model of a RAM with bit-vector address, data-in and data-out ports. The size of the ports should be constrained by generics in the entity interface. Next, develop a test bench that includes a component declaration for the RAM without the generics and with fixed-sized address and data ports. Write a configuration declaration for the test bench that binds the RAM entity to the RAM component instance, using the component local port sizes to determine values for the entity formal generics.

The Predefined Package Standard

The predefined types, subtypes and functions of VHDL are defined in a package called **standard**, stored in the library **std**. Each design unit in a design is automatically preceded by the following context clause:

library std, work; **use** std.standard.**all**;

so the predefined items are directly visible in the design. The package **standard** is listed here. The comments indicate which operators are implicitly defined for each explicitly defined type. These operators are also automatically made visible in design units. The types *universal_integer* and *universal_real* are anonymous types. They cannot be referred to explicitly.

```
package standard is

    type boolean is ( false, true );

        – – implicitly declared for boolean operands:
        – – "and", "or", "nand", "nor", "xor", "xnor", "not"  return boolean
        – – "=", "/=", "<", "<=", ">", ">="  return boolean

    type bit is ( '0', '1' );

        – – implicitly declared for bit operands:
        – – "and", "or", "nand", "nor", "xor", "xnor", "not"  return bit
        – – "=", "/=", "<", "<=", ">", ">="  return boolean
```

type character **is** (

nul,	soh,	stx,	etx,	eot,	enq,	ack,	bel,
bs,	ht,	lf,	vt,	ff,	cr,	so,	si,
dle,	dc1,	dc2,	dc3,	dc4,	nak,	syn,	etb,
can,	em,	sub,	esc,	fsp,	gsp,	rsp,	usp,
' ',	'!',	'"',	'#',	'$',	'%',	'&',	''',
'(',	')',	'*',	'+',	',',	'–',	'.',	'/',
'0',	'1',	'2',	'3',	'4',	'5',	'6',	'7',
'8',	'9',	':',	';',	'<',	'=',	'>',	'?',
'@',	'A',	'B',	'C',	'D',	'E',	'F',	'G',
'H',	'I',	'J',	'K',	'L',	'M',	'N',	'O',
'P',	'Q',	'R',	'S',	'T',	'U',	'V',	'W',
'X',	'Y',	'Z',	'[',	'\',	']',	'^',	'_',
'`',	'a',	'b',	'c',	'd',	'e',	'f',	'g',
'h',	'i',	'j',	'k',	'l',	'm',	'n',	'o',
'p',	'q',	'r',	's',	't',	'u',	'v',	'w',
'x',	'y',	'z',	'{',	'|',	'}',	'~',	del
c128,	c129,	c130,	c131,	c132,	c133,	c134,	c135,
c136,	c137,	c138,	c139,	c140,	c141,	c142,	c143,
c144,	c145,	c146,	c147,	c148,	c149,	c150,	c151,
c152,	c153,	c154,	c155,	c156,	c157,	c158,	c159,
' ',	'¡',	'¢',	'£',	'¤',	'¥',	'¦',	'§'
'¨',	'©',	'ª',	'«',	'¬',	'',	'®',	'¯',
'°',	'±',	'²',	'³',	'´',	'µ',	'¶',	'·',
'¸',	'¹',	'º',	'»',	'¼',	'½',	'¾',	'¿',
'À',	'Á',	'Â',	'Ã',	'Ä',	'Å',	'Æ',	'Ç',
'È',	'É',	'Ê',	'Ë',	'Ì',	'Í',	'Î',	'Ï',
'Ð',	'Ñ',	'Ò',	'Ó',	'Ô',	'Õ',	'Ö',	'×',
'Ø',	'Ù',	'Ú',	'Û',	'Ü',	'Ý',	'Þ',	'ß',
'à',	'á',	'â',	'ã',	'ä',	'å',	'æ',	'ç',
'è',	'é',	'ê',	'ë',	'ì',	'í',	'î',	'ï',
'ð',	'ñ',	'ò',	'ó',	'ô',	'õ',	'ö',	'÷',
'ø',	'ù',	'ú',	'û',	'ü',	'ý',	'þ',	'ÿ');

    – – implicitly declared for character operands:
    – –"=", "/=", "<", "<=", ">", ">=" return boolean

type severity_level **is** (note, warning, error, failure);

    – – implicitly declared for severity_level operands:
    – –"=", "/=", "<", "<=", ">", ">=" return boolean

type *universal_integer* **is range** *implementation defined*;

    – – implicitly declared for universal integer operands:
    – –"=", "/=", "<", "<=", ">", ">=" return boolean
    – –"**", "*", "/", "+", "–", "abs", "rem", "mod" return universal integer

type *universal_real* **is range** *implementation defined*;

    – – implicitly declared for universal real operands:
    – –"=", "/=", "<", "<=", ">", ">=" return boolean
    – –"**", "*", "/", "+", "–", "abs" return universal real

type integer **is range** *implementation defined*;

```
        – – implicitly declared for integer operands:
        – – ”=”, ”/=”, ”<”, ”<=”, ”>”, ”>=”  return boolean
        – – ”**”, ”*”, ”/”, ”+”, ”–”, ”abs”, ”rem”, ”mod”  return integer
```

subtype natural **is** integer **range** 0 **to** integer’high;
subtype positive **is** integer **range** 1 **to** integer’high;

type real **is range** *implementation defined*;

```
        – – implicitly declared for real operands:
        – – ”=”, ”/=”, ”<”, ”<=”, ”>”, ”>=”  return boolean
        – – ”**”, ”*”, ”/”, ”+”, ”–”, ”abs”  return real
```

type time **is range** *implementation defined*
    units
```
            fs;
            ps   = 1000 fs;
            ns   = 1000 ps;
            us   = 1000 ns;
            ms   = 1000 us;
            sec  = 1000 ms;
            min  = 60 sec;
            hr   = 60 min;
```
    end units;

```
        – – implicitly declared for time operands:
        – – ”=”, ”/=”, ”<”, ”<=”, ”>”, ”>=”  return boolean
        – – ”*”, ”+”, ”–”, ”abs”  return time
        – – ”/”  return time or universal integer
```

subtype delay_length **is** time **range** 0 fs **to** time’high;

impure function now **return** delay_length;

type string **is array** (positive **range** <>) **of** character;

```
        – – implicitly declared for string operands:
        – – ”=”, ”/=”, ”<”, ”<=”, ”>”, ”>=”  return boolean
        – – ”&”  return string
```

type bit_vector **is array** (natural **range** <>) **of** bit;

```
        – – implicitly declared for bit_vector operands:
        – – ”and”, ”or”, ”nand”, ”nor”, ”xor”, ”xnor”, ”not”  return bit_vector
        – – ”sll”, ”srl”, ”sla”, ”sra”, ”rol”, ”ror”  return bit_vector
        – – ”=”, ”/=”, ”<”, ”<=”, ”>”, ”>=”  return boolean
        – – ”&”  return bit_vector
```

type file_open_kind **is** (read_mode, write_mode, append_mode);

```
        – – implicitly declared for file_open_kind operands:
        – – ”=”, ”/=”, ”<”, ”<=”, ”>”, ”>=”  return boolean
```

type file_open_status **is** (open_ok, status_error,
                           name_error, mode_error);

```
        – – implicitly declared for file_open_status operands:
        – – ”=”, ”/=”, ”<”, ”<=”, ”>”, ”>=”  return boolean
```

    attribute foreign : string;

end package standard;

VHDL-87

The following items are not included in standard in VHDL-87: the types file_open_kind and file_open_status; the subtype delay_length; the attribute foreign; the operators **xnor**, **sll**, **srl**, **sla**, **sra**, **rol** and **ror**. The result time of the function now is time. The type character includes only the first 128 values, corresponding to the ASCII character set.

IEEE Standard 1164

The IEEE Standard 1164 defines types, operators and functions for detailed modeling of data in a design. It is based on a multivalued logic type, with three states (zero, one and unknown) and three driving strengths (forcing, weak and high-impedance). Forcing strength represents an active driver, such as a transistor or switch connected to a source with approximately zero impedance. Weak strength represents a resistive driver, such as a pull-up resistor or a pass transistor. High-impedance strength represents a driver that is turned off. A value driven with forcing strength dominates a value driven with weak strength, and both dominate a high-impedance driver. This multivalued logic system is augmented with values representing an uninitialized state and a "don't care" state.

The standard-logic package declaration is shown here. The examples in this book show how the operations defined in the package are used. The IEEE standard also includes a package body defining the detailed meaning of each of the operators and functions. However, simulator vendors are allowed to substitute accelerated implementations of the package rather than compiling the package body into a simulation. The IEEE standard requires the package to be in a resource library named **ieee**.

```
package std_logic_1164 is

------------------------------------------------
-- logic state system  (unresolved)
------------------------------------------------

type std_ulogic is ( 'U',    -- Uninitialized
               'X',    -- Forcing  Unknown
               '0',    -- Forcing  0
               '1',    -- Forcing  1
               'Z',    -- High Impedance
               'W',    -- Weak    Unknown
               'L',    -- Weak    0
               'H',    -- Weak    1
               '–'     -- Don't care
          );
```

```
-------------------------------------------------------
-- unconstrained array of std_ulogic for use with the resolution function
-------------------------------------------------------

type std_ulogic_vector is array ( natural range <> ) of std_ulogic;

-------------------------------------------------------
-- resolution function
-------------------------------------------------------

function resolved ( s : std_ulogic_vector ) return std_ulogic;

-------------------------------------------------------
-- *** industry standard logic type ***
-------------------------------------------------------

subtype std_logic is resolved std_ulogic;

-------------------------------------------------------
-- unconstrained array of std_logic for use in declaring signal arrays
-------------------------------------------------------

type std_logic_vector is array ( natural range <> ) of std_logic;

-------------------------------------------------------
-- common subtypes
-------------------------------------------------------

subtype X01    is resolved std_ulogic range 'X' to '1';     -- ('X','0','1')
subtype X01Z   is resolved std_ulogic range 'X' to 'Z';     -- ('X','0','1','Z')
subtype UX01   is resolved std_ulogic range 'U' to '1';     -- ('U','X','0','1')
subtype UX01Z is resolved std_ulogic range 'U' to 'Z';     -- ('U','X','0','1','Z')

-------------------------------------------------------
-- overloaded logical operators
-------------------------------------------------------

function "and"  ( l : std_ulogic; r : std_ulogic ) return UX01;
function "nand" ( l : std_ulogic; r : std_ulogic ) return UX01;
function "or"   ( l : std_ulogic; r : std_ulogic ) return UX01;
function "nor"  ( l : std_ulogic; r : std_ulogic ) return UX01;
function "xor"  ( l : std_ulogic; r : std_ulogic ) return UX01;
function "xnor" ( l : std_ulogic; r : std_ulogic ) return UX01;
function "not"  ( l : std_ulogic               ) return UX01;

-------------------------------------------------------
-- vectorized overloaded logical operators
-------------------------------------------------------

function "and"  ( l, r : std_logic_vector  ) return std_logic_vector;
function "and"  ( l, r : std_ulogic_vector) return std_ulogic_vector;

function "nand" ( l, r : std_logic_vector   ) return std_logic_vector;
function "nand" ( l, r : std_ulogic_vector) return std_ulogic_vector;

function "or"   ( l, r : std_logic_vector  ) return std_logic_vector;
function "or"   ( l, r : std_ulogic_vector) return std_ulogic_vector;

function "nor"  ( l, r : std_logic_vector  ) return std_logic_vector;
function "nor"  ( l, r : std_ulogic_vector) return std_ulogic_vector;
```

```
function "xor"    ( l, r : std_logic_vector  ) return std_logic_vector;
function "xor"    ( l, r : std_ulogic_vector) return std_ulogic_vector;

function "xnor"   ( l, r : std_logic_vector  ) return std_logic_vector;
function "xnor"   ( l, r : std_ulogic_vector) return std_ulogic_vector;

function "not"    ( l : std_logic_vector   ) return std_logic_vector;
function "not"    ( l : std_ulogic_vector  ) return std_ulogic_vector;
```

– – conversion functions

```
function To_bit      ( s : std_ulogic;       xmap : bit := '0' ) return bit;
function To_bitvector ( s : std_logic_vector ; xmap : bit := '0' ) return bit_vector;
function To_bitvector ( s : std_ulogic_vector;xmap : bit := '0' ) return bit_vector;

function To_StdULogic       ( b : bit              ) return std_ulogic;
function To_StdLogicVector  ( b : bit_vector       ) return std_logic_vector;
function To_StdLogicVector  ( s : std_ulogic_vector ) return std_logic_vector;
function To_StdULogicVector ( b : bit_vector        ) return std_ulogic_vector;
function To_StdULogicVector ( s : std_logic_vector  ) return std_ulogic_vector;
```

– – strength strippers and type convertors

```
function To_X01 ( s : std_logic_vector   ) return std_logic_vector;
function To_X01 ( s : std_ulogic_vector ) return std_ulogic_vector;
function To_X01 ( s : std_ulogic         ) return X01;
function To_X01 ( b : bit_vector          ) return std_logic_vector;
function To_X01 ( b : bit_vector          ) return std_ulogic_vector;
function To_X01 ( b : bit                 ) return X01;

function To_X01Z ( s : std_logic_vector   ) return std_logic_vector;
function To_X01Z ( s : std_ulogic_vector ) return std_ulogic_vector;
function To_X01Z ( s : std_ulogic         ) return X01Z;
function To_X01Z ( b : bit_vector          ) return std_logic_vector;
function To_X01Z ( b : bit_vector          ) return std_ulogic_vector;
function To_X01Z ( b : bit                 ) return X01Z;

function To_UX01 ( s : std_logic_vector   ) return std_logic_vector;
function To_UX01 ( s : std_ulogic_vector ) return std_ulogic_vector;
function To_UX01 ( s : std_ulogic         ) return UX01;
function To_UX01 ( b : bit_vector          ) return std_logic_vector;
function To_UX01 ( b : bit_vector          ) return std_ulogic_vector;
function To_UX01 ( b : bit                 ) return UX01;
```

– – edge detection

```
function rising_edge ( signal s : std_ulogic ) return boolean;
function falling_edge ( signal s : std_ulogic ) return boolean;
```

```
-----------------------------------------------
-- object contains an unknown
-----------------------------------------------
function Is_X ( s : std_ulogic_vector ) return boolean;
function Is_X ( s : std_logic_vector ) return boolean;
function Is_X ( s : std_ulogic        ) return boolean;
end std_logic_1164;
```

VHDL-87

The overloaded versions of the **xnor** operator are not included in the VHDL-87 version of the standard-logic package.

VHDL Syntax

In this appendix we present the full set of syntax rules for VHDL using the EBNF notation introduced in Chapter 1. The form of EBNF used in this book differs from that of the *VHDL Language Reference Manual* (*LRM*) in order to make the syntax rules more intelligible to the VHDL user. The *LRM* includes a separate syntax rule for each minor syntactic category. In this book, we condense the grammar into a smaller number of rules, each of which defines a larger part of the grammar. We introduce the EBNF symbols "(", ")" and "₀ ₀" as part of this simplification. Our aim is to avoid the large amount of searching required when using the LRM rules to resolve a question of grammar.

Those parts of the syntax rules that were introduced in VHDL-93 are underlined in this appendix. A VHDL-87 model may not use these features. In addition, there are some entirely new rules, introduced in VHDL-93, that have no predecessors in VHDL-87. We identify these rules individually where they occur in this appendix.

Index to Syntax Rules

C.1   Design File

design_file ⇐ design_unit { ... }
design_unit ⇐
    { library_clause ‖ use_clause }
    library_unit

library_unit ⇐
    entity_declaration | architecture_body
    | package_declaration | package_body
    | configuration_declaration

library_clause ⇐ **library** identifier { , ₀₀₀ } ;

C.2 Library Unit Declarations

entity_declaration ⇐
    entity identifier **is**
        [**generic** (*generic*_interface_list) ;]
        [**port** (*port*_interface_list) ;]
        { entity_declarative_item }
    [**begin**
        { concurrent_assertion_statement
          | *passive*_concurrent_procedure_call_statement
          | *passive*_process_statement }]
    end [**entity**] [identifier] ;

entity_declarative_item ⇐
    subprogram_declaration | subprogram_body
    | type_declaration | subtype_declaration
    | constant_declaration | signal_declaration
    | *shared*_variable_declaration | file_declaration
    | alias_declaration
    | attribute_declaration | attribute_specification
    | disconnection_specification | use_clause
    | group_template_declaration | group_declaration

architecture_body ⇐
    architecture identifier **of** *entity*_name **is**
        { block_declarative_item }
    begin
        { concurrent_statement }
    end [**architecture**] [identifier] ;

configuration_declaration ⇐
    configuration identifier **of** *entity*_name **is**
        { use_clause | attribute_specification | group_declaration }
        block_configuration
    end [**configuration**] [identifier] ;

block_configuration ⇐
    for (*architecture*_name
            | *block_statement*_label
            | *generate_statement*_label ⟦ ((discrete_range | *static*_expression)) ⟧)
       { use_clause }
       { block_configuration
          | **for** component_specification
              ⟦ binding_indication ; ⟧
              ⟦ block_configuration ⟧
          end for ; }
    end for ;

package_declaration ⇐
    package identifier **is**
       { package_declarative_item }
end ⟦ **package** ⟧ ⟦ identifier ⟧ ;

package_declarative_item ⇐
    subprogram_declaration

| type_declaration	| subtype_declaration
| constant_declaration	| signal_declaration
| *shared*_variable_declaration	| file_declaration
| alias_declaration	| component_declaration
| attribute_declaration	| attribute_specification
| disconnection_specification	| use_clause
| group_template_declaration	| group_declaration

package_body ⇐
    package body identifier **is**
       { package_body_declarative_item }
    end ⟦ **package body** ⟧ ⟦ identifier ⟧ ;

package_body_declarative_item ⇐
    subprogram_declaration

subprogram_declaration	| subprogram_body
| type_declaration	| subtype_declaration
| constant_declaration	| *shared*_variable_declaration
| file_declaration	| alias_declaration
| use_clause	
| group_template_declaration	| group_declaration

C.3 Declarations and Specifications

subprogram_specification ⇐
    procedure (identifier | operator_symbol) ⟦ (*parameter*_interface_list) ⟧
    | ⟦ **pure** | **impure** ⟧
     function (identifier | operator_symbol)
         ⟦ (*parameter*_interface_list) ⟧ **return** type_mark
subprogram_declaration ⇐ subprogram_specification ;

subprogram_body ⇐
    subprogram_specification **is**
        { subprogram_declarative_part }
    begin
        { sequential_statement }
    end ⟦ **procedure** ǀ **function** ⟧ ⟦ identifier ǀ operator_symbol ⟧ ;

subprogram_declarative_part ⇐
    subprogram_declaration   ǀ subprogram_body
    ǀ type_declaration   ǀ subtype_declaration
    ǀ constant_declaration   ǀ variable_declaration
    ǀ file_declaration   ǀ alias_declaration
    ǀ attribute_declaration   ǀ attribute_specification
    ǀ use_clause
    ǀ group_template_declaration ǀ group_declaration

type_declaration ⇐
    type identifier **is** type_definition ;
    ǀ **type** identifier ;

type_definition ⇐
    enumeration_type_definition ǀ integer_type_definition
    ǀ floating_type_definition   ǀ physical_type_definition
    ǀ array_type_definition   ǀ record_type_definition
    ǀ access_type_definition   ǀ file_type_definition

constant_declaration ⇐
    constant identifier { , ... } : subtype_indication ⟦ := expression ⟧ ;

signal_declaration ⇐
    signal identifier { , ... } : subtype_indication ⟦ **register** ǀ **bus** ⟧
                                ⟦ := expression ⟧ ;

variable_declaration ⇐
    ⟦ **shared** ⟧ **variable** identifier { , ... } : subtype_indication ⟦ := expression ⟧ ;

file_declaration ⇐
    file identifier { , ... } : subtype_indication
        ⟦ ⟦ **open** *file_open_kind*_expression ⟧ **is** *string*_expression ⟧ ;

VHDL-87

The syntax for a file declaration in VHDL-87 is

file_declaration ⇐
    file identifier : subtype_indication **is** ⟦ **in** ǀ **out** ⟧ *string*_expression ;

This difference is the only case in which VHDL-87 syntax is not a subset of
VHDL-93 syntax. If a VHDL-87 model includes either of the keywords **in** or **out**,
the model cannot be successfully analyzed with a VHDL-93 analyzer.

alias_declaration ⇐
    alias (identifier | character_literal | operator_symbol) [: subtype_indication]
       is name [signature] ;

component_declaration ⇐
    component identifier [**is**]
       [**generic** (*generic*_interface_list) ;]
       [**port** (*port*_interface_list) ;]
    end component [identifier] ;

attribute_declaration ⇐ **attribute** identifier : type_mark ;

attribute_specification ⇐
    attribute identifier **of** entity_name_list : entity_class **is** expression ;

entity_name_list ⇐
    ((identifier | character_literal | operator_symbol) [signature]) { , ... }
    | **others**
    | **all**

entity_class ⇐

entity	**architecture**	**configuration**	**procedure**	**function**
package	**type**	**subtype**	**constant**	**signal**
variable	**component**	**label**		
literal	**units**	**group**	**file**	

configuration_specification ⇐
    for component_specification binding_indication ;

component_specification ⇐
    (*instantiation*_label { , ... } | **others** | **all**) : *component*_name

binding_indication ⇐
    [**use** (**entity** *entity*_name [(*architecture*_identifier)]
           | **configuration** *configuration*_name
           | **open**)]
    [**generic map** (*generic*_association_list)]
    [**port map** (*port*_association_list)]

disconnection_specification ⇐
    disconnect (*signal*_name { , ... } | **others** | **all**) : type_mark
       after *time*_expression ;

group_template_declaration ⇐
    group identifier **is** ((entity_class [<>]) { , ... }) ;

group_declaration ⇐
    group identifier : *group_template*_name ((name | character_literal) { , ... }) ;

VHDL-87

Group template declarations and group declarations are not allowed in VHDL-87.

use_clause ⇐ **use** selected_name { , ∘∘∘ } ;

C.4 Type Definitions

enumeration_type_definition ⇐ ((identifier ∥ character_literal) { , ∘∘∘ })

integer_type_definition ⇐
    range (*range*_attribute_name
          ∥ simple_expression (**to** ∥ **downto**) simple_expression)

floating_type_definition ⇐
    range (*range*_attribute_name
          ∥ simple_expression (**to** ∥ **downto**) simple_expression)

physical_type_definition ⇐
    range (*range*_attribute_name
          ∥ simple_expression (**to** ∥ **downto**) simple_expression)
      units
        identifier ;
        { identifier = physical_literal ; }
      end units ⟦ identifier ⟧

array_type_definition ⇐
    array ((type_mark **range** <>) { , ∘∘∘ }) **of** *element*_subtype_indication
    ∥ **array** (discrete_range { , ∘∘∘ }) **of** *element*_subtype_indication

record_type_definition ⇐
    record
      (identifier { , ∘∘∘ } : subtype_indication ;)
      { ∘∘∘ }
    end record ⟦ identifier ⟧

access_type_definition ⇐ **access** subtype_indication

file_type_definition ⇐ **file of** type_mark

subtype_declaration ⇐ **subtype** identifier **is** subtype_indication ;

subtype_indication ⇐
    ⟦ *resolution_function*_name ⟧
    type_mark
      ⟦ **range** (*range*_attribute_name
          ∥ simple_expression (**to** ∥ **downto**) simple_expression)
      ∥ (discrete_range { , ∘∘∘ }) ⟧

discrete_range ⇐
    *discrete*_subtype_indication
    ∥ *range*_attribute_name
    ∥ simple_expression (**to** ∥ **downto**) simple_expression

type_mark ⇐ *type*_name ∥ *subtype*_name

C.5 Concurrent Statements

concurrent_statement ⇐
    block_statement
    | process_statement
    | concurrent_procedure_call_statement
    | concurrent_assertion_statement
    | concurrent_signal_assignment_statement
    | component_instantiation_statement
    | generate_statement

block_statement ⇐
    *block*_label :
    block [(*guard*_expression)] [**is**]
        [**generic** (*generic*_interface_list) ;
        [**generic map** (*generic*_association_list) ;]]
        [**port** (*port*_interface_list) ;
        [**port map** (*port*_association_list) ;]]
        { block_declarative_item }
    begin
        { concurrent_statement }
    end block [*block*_label] ;

block_declarative_item ⇐
    subprogram_declaration | subprogram_body
    | type_declaration | subtype_declaration
    | constant_declaration | signal_declaration
    | *shared*_variable_declaration | file_declaration
    | alias_declaration | component_declaration
    | attribute_declaration | attribute_specification
    | configuration_specification | disconnection_specification
    | use_clause
    | group_template_declaration | group_declaration

process_statement ⇐
    [*process*_label :]
    [**postponed**] **process** [(*signal*_name { , ... })] [**is**]
        { process_declarative_item }
    begin
        { sequential_statement }
    end [**postponed**] **process** [*process*_label] ;

process_declarative_item ⇐
   subprogram_declaration | subprogram_body
   | type_declaration | subtype_declaration
   | constant_declaration | variable_declaration
   | file_declaration | alias_declaration
   | attribute_declaration | attribute_specification
   | use_clause
   | group_template_declaration | group_declaration

concurrent_procedure_call_statement ⇐
   ⟦ label : ⟧
   ⟦ **postponed** ⟧ *procedure*_name ⟦ (*parameter*_association_list) ⟧ ;

concurrent_assertion_statement ⇐
   ⟦ label : ⟧
   ⟦ **postponed** ⟧ **assert** *boolean*_expression
               ⟦ **report** expression ⟧ ⟦ **severity** expression ⟧ ;

concurrent_signal_assignment_statement ⇐
   ⟦ label : ⟧ ⟦ **postponed** ⟧ conditional_signal_assignment
   | ⟦ label : ⟧ ⟦ **postponed** ⟧ selected_signal_assignment

conditional_signal_assignment ⇐
   (name | aggregate) <=
      ⟦ **guarded** ⟧ ⟦ delay_mechanism ⟧
         { waveform **when** *boolean*_expression **else** }
        waveform ⟦ **when** *boolean*_expression ⟧ ;

selected_signal_assignment ⇐
   with expression **select**
     (name | aggregate) <=
       ⟦ **guarded** ⟧ ⟦ delay_mechanism ⟧
         { waveform **when** choices , }
        waveform **when** choices ;

component_instantiation_statement ⇐
   *instantiation*_label :
      (⟦ **component** ⟧ *component*_name
       | **entity** *entity*_name ⟦ (*architecture*_identifier) ⟧
       | **configuration** *configuration*_name)
     ⟦ **generic map** (*generic*_association_list) ⟧
     ⟦ **port map** (*port*_association_list) ⟧ ;

generate_statement ⇐
   *generate*_label :
   (**for** identifier **in** discrete_range | **if** *boolean*_expression) **generate**
     ⟦ { block_declarative_item }
   begin ⟧
     { concurrent_statement }
   end generate ⟦ *generate*_label ⟧ ;

C.6 Sequential Statements

sequential_statement ⇐

wait_statement		assertion_statement
｜ report_statement		｜ signal_assignment_statement
｜ variable_assignment_statement		｜ procedure_call_statement
｜ if_statement		｜ case_statement
｜ loop_statement		｜ next_statement
｜ exit_statement		｜ return_statement
｜ null_statement		

wait_statement ⇐

⟦ label : ⟧ **wait** ⟦ **on** *signal_*name { , ⚬⚬⚬ } ⟧

⟦ **until** *boolean_*expression ⟧

⟦ **for** *time_*expression ⟧ ;

assertion_statement ⇐

⟦ label : ⟧ **assert** *boolean_*expression

⟦ **report** expression ⟧ ⟦ **severity** expression ⟧ ;

report_statement ⇐ ⟦ label : ⟧ **report** expression ⟦ **severity** expression ⟧ ;

VHDL-87

Report statements are not allowed in VHDL-87.

---

signal_assignment_statement ⇐

⟦ label : ⟧ (name ｜ aggregate) <= ⟦ delay_mechanism ⟧ waveform ;

delay_mechanism ⇐

transport ｜ ⟦ **reject** *time_*expression ⟧ **inertial**

waveform ⇐

(*value_*expression ⟦ **after** *time_*expression ⟧

｜ **null** ⟦ **after** *time_*expression ⟧) { , ⚬⚬⚬ }

｜ **unaffected**

variable_assignment_statement ⇐

⟦ label : ⟧ (name ｜ aggregate) := expression ;

procedure_call_statement ⇐

⟦ label : ⟧ *procedure_*name ⟦ (*parameter_*association_list) ⟧ ;

if_statement ⇐
    ⟦ *if_*label : ⟧
    if *boolean_*expression **then**
        { sequential_statement }
    { **elsif** *boolean_*expression **then**
        { sequential_statement } }
    ⟦ **else**
        { sequential_statement } ⟧
    end if ⟦ *if_*label ⟧ ;

case_statement ⇐
    ⟦ *case_*label : ⟧
    case expression **is**
        (**when** choices => { sequential_statement })
        { ∘∘∘ }
    end case ⟦ *case_*label ⟧ ;

loop_statement ⇐
    ⟦ *loop_*label : ⟧
    ⟦ **while** condition ⎮ **for** identifier **in** discrete_range ⟧ **loop**
        { sequential_statement }
    end loop ⟦ *loop_*label ⟧ ;

next_statement ⇐ ⟦ label : ⟧ **next** ⟦ *loop_*label ⟧ ⟦ **when** *boolean_*expression ⟧ ;

exit_statement ⇐ ⟦ label : ⟧ **exit** ⟦ *loop_*label ⟧ ⟦ **when** *boolean_*expression ⟧ ;

return_statement ⇐ ⟦ label : ⟧ **return** ⟦ expression ⟧ ;

null_statement ⇐ ⟦ label : ⟧ **null** ;

C.7   Interfaces and Associations

interface_list ⇐
    (⟦ **constant** ⟧ identifier { , ∘∘∘ } : ⟦ **in** ⟧ subtype_indication
                                     ⟦ := *static_*expression ⟧
    ⎮ ⟦ **signal** ⟧ identifier { , ∘∘∘ } : ⟦ mode ⟧ subtype_indication ⟦ **bus** ⟧
                                       ⟦ := *static_*expression ⟧
    ⎮ ⟦ **variable** ⟧ identifier { , ∘∘∘ } : ⟦ mode ⟧ subtype_indication
                                       ⟦ := *static_*expression ⟧
    ⎮ **file** identifier { , ∘∘∘ } : subtype_indication) { ; ∘∘∘ }

mode ⇐ **in** ⎮ **out** ⎮ **inout** ⎮ **buffer** ⎮ **linkage**

association_list ⇐ (⟦ formal_part => ⟧ actual_part) { , ∘∘∘ }

formal_part ⇐
    *generic_*name
    ⎮ *port_*name
    ⎮ *parameter_*name
    ⎮ *function_*name ((*generic_*name ⎮ *port_*name ⎮ *parameter_*name))
    ⎮ type_mark ((*generic_*name ⎮ *port_*name ⎮ *parameter_*name))

actual_part ⇐
    expression
    | *signal*_name
    | *variable*_name
    | **open**
    | *function*_name ((*signal*_name | *variable*_name))
    | type_mark ((*signal*_name | *variable*_name))

C.8 Expressions

expression ⇐
    relation { **and** relation } | relation [**nand** relation]
    | relation { **or** relation } | relation [**nor** relation]
    | relation { **xor** relation } | relation { **xnor** relation }

relation ⇐ shift_expression [(= | /= | < | <= | > | >=) shift_expression]

VHDL-87

The syntax for a relation in VHDL-87 is

relation ⇐
    simple_expression [(= | /= | < | <= | > | >=) simple_expression]

shift_expression ⇐
    simple_expression [(**sll** | **srl** | **sla** | **sra** | **rol** | **ror**) simple_expression]

simple_expression ⇐ [+ | −] term { (+ | − | &) term }

term ⇐ factor { (* | / | **mod** | **rem**) factor }

factor ⇐ primary [** primary] | **abs** primary | **not** primary

primary ⇐
    name | literal
    | aggregate | function_call
    | qualified_expression | type_mark (expression)
    | **new** subtype_indication | **new** qualified_expression
    | (expression)

function_call ⇐ *function*_name [(*parameter*_association_list)]

qualified_expression ⇐ type_mark ' (expression) | type_mark ' aggregate

name ⇐
    identifier
    | operator_symbol
    | selected_name
    | (name | function_call) (expression { , ... })
    | (name | function_call) (discrete_range)
    | attribute_name

selected_name ⇐
    (name ‖ function_call) . (identifier ‖ character_literal ‖ operator_symbol ‖ **all**)

operator_symbol ⇐ " { graphic_character } "

attribute_name ⇐ (name ‖ function_call) ⟦ signature ⟧ ' identifier ⟦ (expression) ⟧

signature ⇐ ⟦ ⟦ type_mark { , ₒₒₒ } ⟧ ⟧ ⟦ **return** type_mark ⟧ ⟧

literal ⇐
    decimal_literal
    ‖ based_literal
    ‖ ⟦ decimal_literal ‖ based_literal ⟧ *unit*_name
    ‖ identifier
    ‖ character_literal
    ‖ string_literal
    ‖ bit_string_literal
    ‖ **null**

decimal_literal ⇐ integer ⟦ . integer ⟧ ⟦ E ⟦ + ⟧ integer ‖ E − integer ⟧

based_literal ⇐
    integer # based_integer ⟦ . based_integer ⟧ # ⟦ E ⟦ + ⟧ integer ‖ E − integer ⟧

integer ⇐ digit { ⟦ _ ⟧ ₒₒₒ }

based_integer ⇐ (digit ‖ letter) { ⟦ _ ⟧ ₒₒₒ }

character_literal ⇐ ' graphic_character '

string_literal ⇐ " { graphic_character } "

bit_string_literal ⇐ (B ‖ O ‖ X) " ⟦ (digit ‖ letter) { ⟦ _ ⟧ ₒₒₒ } ⟧ "

aggregate ⇐ ((⟦ choices => ⟧ expression) { , ₒₒₒ })

choices ⇐ (simple_expression ‖ discrete_range ‖ identifier ‖ **others**) { ‖ ₒₒₒ }

label ⇐ identifier

identifier ⇐ letter { ⟦ _ ⟧ (letter ‖ digit) } ‖ \ graphic_character { ₒₒₒ } \

Differences Between VHDL-87 and VHDL-93

In this appendix we summarize the restrictions of VHDL-87 that we mentioned throughout the book. We take the restrictive approach, rather than describing the way in which VHDL-93 is an extension of VHDL-87, since VHDL-93 is now the "official" version of the language. We expect most new models to be written in VHDL-93. Nevertheless, designers must deal with the legacy of models written before adoption of the new version of the language and with VHDL tools that have yet to be updated to handle the new version.

Lexical Differences

VHDL-87 uses the ASCII character set, rather than the full ISO character set. ASCII is a subset of the ISO character set, consisting of just the first 128 characters. This includes all of the unaccented letters, but excludes letters with diacritical marks.

VHDL-87 only allows basic identifiers, not extended identifiers. The rules for forming basic identifiers are the same as those for VHDL-93.

The following identifiers are not used as reserved words in VHDL-87. They may be used as identifiers for other purposes, although it is not advisable to do so, as this may cause difficulties in porting the models to VHDL-93.

group	**postponed**	**ror**	**sra**
impure	**pure**	**shared**	**srl**
inertial	**reject**	**sla**	**unaffected**
literal	**rol**	**sll**	**xnor**

Bit-string literals may only be used as literals for array types in which the elements are of type bit. The predefined type bit_vector is such a type. However, the standard-logic types std_ulogic_vector and std_logic_vector are not.

Syntactic Differences

The only sequential statements that may be labeled in VHDL-87 are loop statements. The remaining sequential statements, which may not be labeled, are variable assignment, signal assignment, wait, if, case, null, exit, next, assertion, procedure call and return statements.

The keyword **entity** may not be repeated at the end of an entity declaration.

The keyword **architecture** may not be repeated at the end of an architecture body.

The keyword **configuration** may not be repeated at the end of a configuration declaration.

The keyword **package** may not be repeated at the end of a package declaration.

The keywords **package body** may not be repeated at the end of a package body.

The keyword **procedure** may not be repeated at the end of a procedure declaration.

The keyword **function** may not be repeated at the end of a function declaration.

In a physical type definition, the type name may not be repeated after the keywords **end units**.

In a record type definition, the type name may not be repeated after the keywords **end record**.

The keyword **is** may not be included in the header of a block statement.

The keyword **is** may not be included in the header of a process statement.

The keyword **is** may not be included in the header of a component declaration, and the component name may not be repeated at the end of the declaration.

A generate statement may not include a declarative part or the keyword **begin**.

A component instantiation statement may not directly instantiate an entity or a configuration. It may only instantiate a declared component, but may not include the keyword **component**.

A conditional signal assignment statement may not include the keyword **else** and a condition after the last waveform in the statement.

The keyword **unaffected** may not be used in conditional and selected signal assignment statements.

An alias declaration in VHDL-87 must include a subtype indication.

The VHDL-87 syntax for file declarations is not a subset of the VHDL-93 syntax. The syntax rule in VHDL-87 is

>   file_declaration ⇐
>   **file** identifier : subtype_indication **is**
>   〚 **in** ‖ **out** 〛 *string*_expression ;

An attribute specification may not name a character literal as an item to be decorated, nor specify the entity class **literal**, **units**, **group** or **file**. An attribute specification may not include a signature after an item name.

Semantic Differences

In VHDL-87, the range specified in a slice may have the opposite direction to that of the index range of the array. In this case, the slice is a null slice.

The VHDL-87 language definition does not disallow the keyword **bus** in the specification of a signal parameter. However, it does not specify whether the kind of signal, guarded or unguarded, is determined by the formal parameter specification or by the actual signal associated with the parameter. Implementations of VHDL-87 make different interpretations. Some require the formal parameter specification to include the keyword **bus** if the procedure includes a null signal assignment to the parameter. The actual signal associated with the parameter in a procedure call must then be a guarded signal. Other implementations follow the approach adopted in VHDL-93, prohibiting the keyword **bus** in the parameter specification and determining the kind of the parameter from the kind of the actual signal.

In VHDL-87, files are of the variable class of objects. Hence file parameters in subprograms are specified as variable-class parameters. A subprogram that reads a file parameter should declare the parameter to be of mode **in**. A subprogram that writes a file parameter should declare the parameter to be of mode **out**.

Differences in the Standard Environment

The types file_open_kind and file_open_status, the subtype delay_length and the attribute foreign are not declared in std.standard. The function now returns a value of type time.

Since VHDL-87 uses the ASCII character set, the type character includes only the 128 ASCII characters.

The predefined attributes 'ascending, 'image, 'value, 'driving, 'driving_value, 'simple_name, 'path_name and 'instance_name are not provided.

The logical operator **xnor** and the shift operators **sll**, **srl**, **sla**, **sra**, **rol** and **ror** are not provided and so cannot be declared as overloaded operators. Thus, the VHDL-87 version of the standard-logic package std_logic_1164 does not define the **xnor** operator for standard-logic types.

In VHDL-87, the 'last_value attribute for a composite signal returns the aggregate of last values for each of the scalar elements of the signal, as opposed to the last value of the entire composite signal.

The VHDL-87 version of the textio package declares the function endline, which is not included in the VHDL-93 version of the package.

VHDL-93 Facilities Not in VHDL-87

VHDL-87 does not include report statements, labeled sequential statements (and hence decoration of sequential statements with attributes), postponed processes, shared variables, group templates or groups, aliases for non-data objects, declarative parts in generate statements or file open and close operations.

VHDL-87 does not allow specification of a pulse rejection interval in the delay mechanism part of a signal assignment. Transport delay can be specified using the keyword **transport**. If it is omitted, inertial delay is assumed, with a pulse rejection interval equal to the inertial delay interval.

VHDL-87 does not allow association of an expression with a port in a port map. VHDL-87 does not allow type conversions in association lists.

VHDL-87 does not allow direct instantiation of an entity or a configuration. Only declared components can be instantiated. They must be bound to design entities using configuration specifications or using component configurations in configuration declarations.

VHDL-87 does not allow incremental binding. It is an error if a design includes both a configuration specification and a component configuration for a given component instance.

Answers to Exercises

In this Appendix, we provide sample answers to the quiz-style exercises marked with the symbol "❶". Readers are encouraged to test their answers to the other, more involved, exercises by running the models on a VHDL simulator.

Chapter 1

1. Entity declaration: defines the interface to a module, in terms of its ports, their data transfer direction and their types. Behavioral architecture body: defines the function of a module in terms of an algorithm. Structural architecture body: defines an implementation of a module in terms of an interconnected composition of submodules. Process statement: encapsulates an algorithm in a behavioral description, contains sequential actions to be performed. Signal assignment statement: specifies values to be applied to signals at some later time. Port map: specifies the interconnection between signals and component instance ports in a structural architecture.

2.
```
apply_transform : process is
begin
    d_out <= transform(d_in) after 200 ps;
    -- debug_test <= transform(d_in);
    wait on enable, d_in;
end process apply_transform;
```

3. Basic identifiers: last_item. Reserved words: buffer. Invalid: prev item, value–1 and element#5 include characters that may not occur within identifiers; _control starts with an underscore; 93_999 starts with a digit; entry_ ends with an underscore.

4.   16#1#   16#22#   16#100.0#   16#0.8#

5.   12   132   44   250000   32768   0.625

6. The literal 16#23DF# is an integer expressed in base 16, whereas the literal X"23DF" is a string of 16 bits.

7.   B"111_100_111"   B"011_111_111"   B"001_011_100_101"
      B"1111_0010"   B"0000_0000_0001_0100"
      B"0000_0000_0000_0000_0000_0000_0000_0001"

Chapter 2

1.   **constant** bits_per_word : integer := 32;
      constant pi : real := 3.14159;

2.   **variable** counter : integer := 0;
      variable busy_status : boolean;
      variable temp_result : std_ulogic;

3.   counter := counter + 1;
      busy_status := true;
      temp_result := 'W';

4.   **package** misc_types **is**
          type small_int **is range** 0 **to** 255;
          type fraction **is range** −1.0 **to** +1.0;
          type current **is range** integer'low **to** integer'high
              units nA;
                  uA = 1000 nA;
                  mA = 1000 uA;
                  A = 1000 mA;
              end units;
          type colors **is** (red, yellow, green);
      end package misc_types;

5.   pulse_range'left = pulse_range'low = 1 ms
      pulse_range'right = pulse_range'high = 100 ms
      pulse_range'ascending = true
      word_index'left = 31   word_index'right = 0
      word_index'low = 0   word_index'high = 31
      word_index'ascending = false

6.   state'pos(standby) = 1   state'val(2) = active1
      state'succ(active2) is undefined   state'pred(active1) = standby
      state'leftof(off) is undefined   state'rightof(off) = standby

7.   2 * 3 + 6 / 4 = 7
      3 + −4 is syntactically incorrect
      "cat" & character'('0') = "cat0"
      true **and** x **and not** y **or** z is syntactically incorrect
      B"101110" **sll** 3= B"000101"
      B"100010" **sra** 2 & X"2C"= B"11100000111100"

Chapter 3

1.  ```
    if n mod 2 = 1 then
        odd := '1';
    else
        odd := '0';
    end if;
    ```

2.  ```
    if year mod 100 = 0 then
        days_in_February := 28;
    elsif year mod 4 = 0 then
        days_in_February := 29;
    else
        days_in_February := 28;
    end if;
    ```

3.  ```
    case x is
        when '0' | 'L' => x := '0';
        when '1' | 'H' => x := '1';
        when others => x := 'X';
    end case;
    ```

4.  ```
    case ch is
        when 'A' to 'Z' | 'a' to 'z' | 'À' to 'Ö' | 'Ø' to 'ß' | 'à' to 'ö' | 'ø' to 'ÿ' =>
            character_class := 1;
        when '0' to '9' =>  character_class := 2;
        when nul to usp | del | c128 to c159 =>  character_class := 4;
        when others => character_class := 3;
    end case;
    ```

5.  ```
    loop
        wait until clk = '1';
        exit when d = '1';
    end loop;
    ```

6.  ```
    sum := 1.0;
    term := 1.0;
    n := 0;
    while abs term > abs (sum / 1.0E5) loop
        n := n + 1;
        term := term * x / real(n);
        sum := sum + term;
    end loop;
    ```

7.  ```
    sum := 1.0;
    term := 1.0;
    for n in 1 to 7 loop
        term := term * x / real(n);
        sum := sum + term;
    end loop;
    ```

8.  ```
    assert to_X01(q) = not to_X01(q_n)
        report "flipflop outputs are not complementary";
    ```

9. Insert the statement after the comment "*– – at this point, reset = '1'*":

    ```
    report "counter is reset";
    ```

Chapter 4

1. **type** num_vector **is array** (1 **to** 30) **of** integer;
    variable numbers : num_vector;

    . . .
    sum := 0;
    for i **in** numbers'range **loop**
        sum := sum + numbers(i);
    end loop;
    average := sum / numbers'length;

2. **type** std_ulogic_to_bit_array **is array** (std_ulogic) **of** bit;
    constant std_ulogic_to_bit : std_ulogic_to_bit_array
        := ('U' => '0', 'X' => '0', '0' => '0', '1' => '1', 'Z' => '0',
            'W' => '0', 'L' => '0', 'H' => '1', '–' => '0');

    . . .
    for index **in** 0 **to** 15 **loop**
        v2(index) := std_ulogic_to_bit(v1(index));
    end loop;

3. **type** free_map_array **is array** (0 **to** 1, 0 **to** 79, 0 **to** 17) **of** bit;
    variable free_map : free_map_array;

    . . .
    found := false;
    for side **in** 0 **to** 1 **loop**
        for track **in** 0 **to** 79 **loop**
            for sector **in** 0 **to** 17 **loop**
                if free_map(side, track, sector) = '1' **then**
                    found := true; free_side := side;
                    free_track := track; free_sector := sector;
                    exit;
                end if;
            end loop;
        end loop;
    end loop;

4. **subtype** std_ulogic_byte **is** std_ulogic_vector(7 **downto** 0);
    constant Z_byte : std_ulogic_byte := "ZZZZZZZZ";

5. count := 0;
    for index **in** v'range **loop**
        if v(index) = '1' **then**
            count := count + 1;
        end if;
    end loop;

6. Assuming the declarations

    variable v1 : bit_vector(7 **downto** 0);
    variable v2 : bit_vector(31 **downto** 0);

    . . .
    v2(31 **downto** 24) := v1;
    v2 := v2 **sra** 24;

7. **type** test_record **is record**
            stimulus : bit_vector(0 **to** 2);

```
        delay : delay_length;
        expected_response : bit_vector(0 to 7);
    end record test_record;
```

Chapter 5

1.
```
    entity lookup_ROM is
        port ( address : in lookup_index;  data : out real );

        type lookup_table is array (lookup_index) of real;
        constant lookup_data : lookup_table
            := ( real'high, 1.0, 1.0/2.0, 1.0/3.0, 1.0/4.0, . . . );

    end entity lookup_ROM;
```

2. Transactions are 'Z' at 0 ns, '0' at 10 ns, '1' at 30 ns, '1' at 55 ns, 'H' at 65 ns and 'Z' at 100 ns. The signal is active at all of these times. Events occur at each time except 55 ns, since the signal already has the value '1' at that time.

3. s'delayed(5 ns): 'Z' at 5 ns, '0' at 15 ns, '1' at 35 ns, 'H' at 70 ns, 'Z' at 105 ns. s'stable(5 ns): false at 0 ns, true at 5 ns, false at 10 ns, true at 15 ns, false at 30 ns, true at 35 ns, false at 65 ns, true at 70 ns, false at 100 ns, true at 105 ns. s'quiet(5 ns): false at 0 ns, true at 5 ns, false at 10 ns, true at 15 ns, false at 30 ns, true at 35 ns, false at 55 ns, true at 60 ns, false at 65 ns, true at 70 ns, false at 100 ns, true at 105 ns. s'transaction (assuming an initial value of '0'): '1' at 0 ns, '0' at 10 ns, '1' at 30 ns, '0' at 55 ns, '1' at 65 ns, '0' at 100 ns. At time 60 ns, s'last_event is 30 ns, s'last_active is 5 ns, and s'last_value is '0'.

4.
```
    wait on s until s = '1' and en = '1';
```

5.
```
    wait until ready = '1' for 5 ms;
```

6. The variable v1 is assigned false, since s is not updated until the next simulation cycle. The variable v2 is assigned true, since the wait statement causes the process to resume after s is updated with the value '1'.

7. At 0 ns: schedule '1' for 6 ns. At 3 ns: schedule '0' for 7 ns. At 8 ns: schedule '1' for 14 ns. At 9 ns: delete transaction scheduled for 14 ns, schedule '0' for 13 ns. The signal z takes on the values '1' at 6 ns and '0' at 7 ns. The transaction scheduled for 13 ns does not result in an event on z.

8. At 0 ns: schedule 1 for 7 ns, 23 for 9 ns, 5 for 10 ns, 23 for 12 ns and –5 for 15 ns. At 6 ns: schedule 23 for 13 ns, delete transactions scheduled for 15 ns, 10 ns and 9 ns. The signal x takes on the values 1 at 7 ns and 23 at 12 ns.

9.
```
    mux_logic : process is
    begin
        if enable = '1' and sel = '0' then
            z <= a and not b after 5 ns;
        elsif enable = '1' and sel = '1' then
            z <= x or y after 6 ns;
        else
            z <= '0' after 4 ns;
        end if;
        wait on a, b, enable, sel, x, y;
    end process mux_logic;
```

10.   **process is**
         begin
                 case bit_vector'(s, r) **is**
                     when "00" => **null**;
                     when "01" => q <= '0';
                     when "10" | "11" => q <= '1';
                 end case;
                 wait on s, r;
         end process;

11.   **assert** clk'last_event >= T_pw_clk
             report "interval between changes on clk is too small";

12.   bit_0 : **entity** work.ttl_74x74(basic)
             port map (pr_n => '1', d => q0_n, clk => clk, clr_n => reset,
                       q => q0, q_n => q0_n);

         bit_1 : **entity** work.ttl_74x74(basic)
             port map (pr_n => '1', d => q1_n, clk => q0_n, clr_n => reset,
                       q => q1, q_n => q1_n);

13.

14. One possible order is suggested: analyzing all entity declarations first, followed
    by all architecture bodies:

         entity edge_triggered_Dff
         entity reg4
         entity add_1
         entity buf4
         entity counter
         architecture behav of edge_triggered_Dff

```
architecture struct of reg4
architecture boolean_eqn of add_1
architecture basic of buf
architecture registered of counter
```

An alternative is

```
entity counter
entity buf4
entity add_1
entity reg4
architecture registered of counter
architecture basic of buf
architecture boolean_eqn of add_1
entity edge_triggered_Dff
architecture struct of reg4
architecture behav of edge_triggered_Dff
```

15. **library** company_lib, project_lib;
    use company_lib.in_pad, company_lib.out_pad, project_lib.**all**;

Chapter 6

1. **constant** operand1 : **in** integer
   operand1 : integer
   constant tag : **in** bit_vector(31 **downto** 16)
   tag : bit_vector(31 **downto** 16)
   constant trace : **in** boolean := false
   trace : boolean := false

2. **variable** average : **out** real;
   average : **out** real
   variable identifier : **inout** string
   identifier : **inout** string

3. **signal** clk : **out** bit
   signal data_in : **in** std_ulogic_vector
   signal data_in : std_ulogic_vector

4. Some alternatives are

   ```
   stimulate ( s, 5 ns, 3 );
   stimulate ( target => s, delay => 5 ns, cycles => 3 );

   stimulate ( s, 10 ns, 1 );
   stimulate ( s, 10 ns );
   stimulate ( target => s, delay => 10 ns, cycles => open );
   stimulate ( target => s, cycles => open, delay => 10 ns );
   stimulate ( target => s, delay => 10 ns );

   stimulate ( s, 1 ns, 15 );
   stimulate ( target => s, delay => open, cycles => 15 );
   stimulate ( target => s, cycles => 15);
   stimulate ( s, cycles => 15);
   ```

5.   swapper : **process is**
       begin
           shuffle_bytes (ext_data, int_data, swap_control, Tpd_swap);
           wait on ext_data, swap_control;
       end process swapper;

6.   product_size := approx_log_2(multiplicand) + approx_log_2(multiplier);

7.   **assert** now <= 20 ms
           report "simulation time has exceeded 20 ms";

8. The third, first, none and third, respectively.

9.   **architecture** behavioral **of** computer system **is**

           signal internal_data : bit_vector(31 **downto** 0);

           interpreter : **process is**

               variable opcode : bit_vector(5 **downto** 0);

               procedure do_write **is**
                   variable aligned_address : natural;
               begin
                   . . .
               end procedure do_write;

           begin

               . . .

           end process interpreter;

       end architecture behavioral;

Chapter 7

1.   **package** EMS_types **is**
           type engine_speed **is range** 0 **to** integer'high
               units rpm;
               end units engine_speed;
           constant peak_rpm : engine_speed := 6000 rpm;
           type gear **is** (first, second, third, fourth, reverse);
       end package EMS_types;

       work.EMS_types.engine_speed
       work.EMS_types.rpm   work.EMS_types.peak_rpm
       work.EMS_types.gear   work.EMS_types.first
       work.EMS_types.second   work.EMS_types.third
       work.EMS_types.fourth   work.EMS_types.reverse

2.   **procedure** increment (num : **inout** integer);

3.   **function** odd (num : integer) **return** boolean;

4.   **constant** e : real;

5. No. The package does not contain any subprogram declarations or deferred
   constant declarations.

6.   **use** work.EMS_types.engine_speed;

7.   **library** DSP_lib;
     use DSP_lib.systolic_FFT, DSP_lib.DSP_types.**all**;

Chapter 8

1. (a) '1'.

   (b) '0'.

   (c) Either '1' or '0'. The order of contributions within the array passed to the resolution function is not defined. This particular resolution function returns the leftmost non-'Z' value in the array, so the result depends on the order in which the simulator assembles the contributions.

2.   **subtype** wired_and_logic **is** wired_and tri_state_logic;
     signal synch_control : wired_and_logic := '0';

3. The initial value is 'X'. The default initial value of type **MVL4**, 'X', is used as the initial value of each driver of **int_req**. These contributions are passed to the resolution function, which returns the value 'X'.

4. No, since the operation represented by the table in the resolution function is commutative and associative, with 'Z' as its identity.

5. (a) "ZZZZ0011"

   (b) "XXXX0011"

   (c) "0011XX11"

6. "XXXXZZZZ00111100"

7. (a) '0'

   (b) '0'

   (c) 'W'

   (d) 'U'

   (e) 'X'

8.

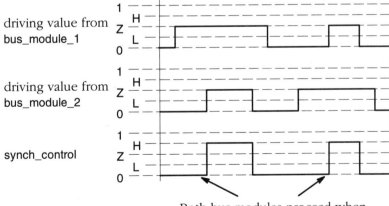

Both bus modules proceed when synch_control changes to 'H'

9. The resolution function is invoked seven times: for the Mem port, the Cache port, the CPU/Mem Section port, the Serial port, the DMA port, the I/O Section port and the Data Bus signal.

10. We cannot simply invert the value read from the port, since the value may differ from that driven by the process. Instead, we use the 'driving_value attribute:

    synch_T <= **not** synch_T'driving_value;

Chapter 9

1.
```
entity flipflop is
    generic ( Tpw_clk_h, T_pw_clk_l : delay_length := 3 ns );
    port ( clk, d : in bit;  q, q_n : out bit );
end entity flipflop;
```

2.
```
clk_gen : entity work.clock_generator
    generic map ( period => 10 ns )
    port map ( clk => master_clk );
```

3.
```
entity adder is
    generic ( data_length : positive );
    port ( a, b : in std_logic_vector(data_length – 1 downto 0);
           sum : out std_logic_vector(data_length – 1 downto 0) );
end entity adder;
```

4.
```
io_control_reg : entity work.reg
    generic map ( width => 4 )
    port map ( d => data_out(3 downto 0),
               q(0) => io_en, q(1) => io_int_en, q(2) => io_dir, q(3) => io_mode,
               clk => io_write, reset => io_reset );
```

Chapter 10

1. An entity declaration uses the keyword **entity** where a component declaration uses the keyword **component**. An entity declaration is a design unit that is analyzed and placed into a design library, whereas a component declaration is simply a declaration in an architecture body or a package. An entity declaration has a declarative part and a statement part, providing part of the implementation of the interface, whereas a component declaration simply declares an interface with no implementation information. An entity declaration represents the interface of a "real" electronic circuit, whereas a component declaration represents a "virtual" or "template" interface.

2.
```
component magnitude_comparator is
    generic ( width : positive; Tpd : delay_length );
    port ( a, b : in std_logic_vector(width – 1 downto 0);
           a_equals_b, a_less_than_b : out std_logic );
end component magnitude_comparator;
```

3.
```
position_comparator : component magnitude_comparator
    generic map ( width => current_position'length, Tpd => 12 ns )
    port map ( a => current_position, b => upper_limit,
               a_less_than_b => position_ok, a_equals_b => open );
```

4.
```
package small_number_pkg is
    subtype small_number is natural range 0 to 255;
    component adder is
        port ( a, b : in small_number;  s : out small_number );
    end component adder;
end package small_number_pkg;
```

5.
```
library dsp_lib;
configuration digital_filter_rtl of digital_filter is
    for register_transfer
        for coeff_1_multiplier : multiplier
            use entity dsp_lib.fixed_point_mult(algorithmic);
        end for;
    end for;
end configuration digital_filter_rtl;
```

6.
```
library dsp_lib;
configuration digital_filter_std_cell of digital_filter is
    for register_transfer
        for coeff_1_multiplier : multiplier
            use configuration dsp_lib.fixed_point_mult_std_cell;
        end for;
    end for;
end configuration digital_filter_std_cell;
```

7.
```
library dsp_lib;
architecture register_transfer of digital_filter is
    . . .
begin
    coeff_1_multiplier : configuration dsp_lib.fixed_point_mult_std_cell
        port map ( . . . );
    . . .
end architecture register_transfer;
```

8.
```
use entity work.multiplexer
    generic map ( Tpd => 3.5 ns );
```

9.

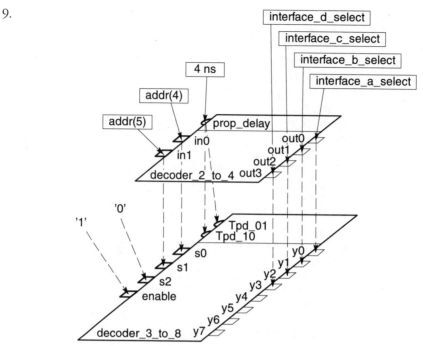

10.   **generic map** (Tpd_01 => **open**, Tpd_10 => **open**)
      port map (a => a, b => b, c => c, d => **open**, y => y)

References

[1] C. G. Bell and A. Newell, *Computer Structures: Readings and Examples*, McGraw-Hill, New York, 1971.

[2] D. D. Gajski and R. H. Kuhn, "New VLSI Tools," *IEEE Computer*, Vol. 16, no. 12 (December 1983), pp. 11-14.

Index